Developments in Geotechnical Engineering

Series editor
Braja M. Das, Henderson, USA
Nagaratnam Sivakugan, Townsville, Australia

More information about this series at http://www.springer.com/series/13410

Jay Ameratunga · Nagaratnam Sivakugan
Braja M. Das

Correlations of Soil and Rock Properties in Geotechnical Engineering

Jay Ameratunga
Golder Associates
Brisbane, QLD, Australia

Nagaratnam Sivakugan
James Cook University
Townsville, QLD, Australia

Braja M. Das
California State University
Henderson, CA, USA

ISSN 2364-5156 ISSN 2364-5164 (electronic)
Developments in Geotechnical Engineering
ISBN 978-81-322-2627-7 ISBN 978-81-322-2629-1 (eBook)
DOI 10.1007/978-81-322-2629-1

Library of Congress Control Number: 2015951662

Springer New Delhi Heidelberg New York Dordrecht London
© Springer India 2016
This work is subject to copyright. All rights are reserved by the Publisher, whether the whole or part of the material is concerned, specifically the rights of translation, reprinting, reuse of illustrations, recitation, broadcasting, reproduction on microfilms or in any other physical way, and transmission or information storage and retrieval, electronic adaptation, computer software, or by similar or dissimilar methodology now known or hereafter developed.
The use of general descriptive names, registered names, trademarks, service marks, etc. in this publication does not imply, even in the absence of a specific statement, that such names are exempt from the relevant protective laws and regulations and therefore free for general use.
The publisher, the authors and the editors are safe to assume that the advice and information in this book are believed to be true and accurate at the date of publication. Neither the publisher nor the authors or the editors give a warranty, express or implied, with respect to the material contained herein or for any errors or omissions that may have been made.

Printed on acid-free paper

Springer (India) Pvt. Ltd. is part of Springer Science+Business Media (www.springer.com)

Preface

Geotechnical engineering has grown rapidly in the past half century with the contribution from academics, researchers and practising professionals. It is still considered a combination of art and science with research and observations in the field refining and improving geotechnical design. Although in situ and laboratory geotechnical testing still remain the two preferred methods of determining design parameters, empiricism has a unique and a big role to play in geotechnical engineering.

Geotechnical literature is full of empirical equations and graphs, and they are used regularly by practitioners worldwide. These are derived based on laboratory or field data, past experience and good judgement. Where little or no geotechnical information is available, or where reasonableness of a test result needs to be checked, these empirical equations provide an alternative very useful to the engineer. For some parameters, you may be confronted with several empirical equations, and it is a good practice to clearly state the source so that the readers can make their own judgement.

The main objective of this book is to provide correlations commonly used by geotechnical practitioners to assess design parameters important in the geotechnical design activities. It is intended mainly for the practitioners although its value extends to academics and researchers as well. We have arranged the chapters on the basis of the main types of in situ tests with laboratory tests on soil and rock given two separate chapters. In Chapter 2, we have provided a brief overview of the geotechnical properties commonly determined in the laboratory, their relevance in soil mechanics and laboratory tests for determining them. It gives the necessary background for the chapters that follow.

Acknowledgements

We are thankful to Professor Harry Poulos who provided valuable suggestions on the first draft. Thanks are also due to Dr. Chris Bridges and Thayalan Nallarulanatham for assisting and reviewing some of the chapters. Allan McConnell of Insitu Geotech Services and Yvo Keulemans of Cone Penetration Testing Services have provided some valuable pictures for this book. We are thankful to them for their contributions. Finally we would like to thank the staff at Springer for their assistance and advice, especially Swati Meherishi, senior editor, and Kamiya Khatter, senior editorial assistant.

Brisbane, QLD, Australia	Jay Ameratunga
Townsville, QLD, Australia	Nagaratnam Sivakugan
Henderson, CA, USA	Braja M. Das

Contents

1	**Introduction**		1
	1.1	Laboratory Testing	1
	1.2	In Situ Testing	2
	1.3	Empirical Correlations	5
	1.4	Contents of the Book	8
	References		9
2	**Geotechnical Properties of Soils – Fundamentals**		11
	2.1	Laboratory Tests for Soils	11
	2.2	Phase Relations	12
		2.2.1 Terminology and Definitions	13
		2.2.2 Relationships Between the Variables	15
	2.3	Granular Soils	16
		2.3.1 Grain Size Distribution	16
		2.3.2 Relative Density	18
	2.4	Plasticity	19
		2.4.1 Atterberg Limits	19
		2.4.2 Classification of Fine Grained Soils Based on Plasticity	20
	2.5	Compaction	21
	2.6	Permeability	24
		2.6.1 D'Arcy's Law and Permeability Measurements	24
		2.6.2 Intrinsic Permeability	28
		2.6.3 Reynold's Number and Laminar Flow	29
		2.6.4 Anisotropy	29
		2.6.5 One-Dimensional Flow in Layered Soils	30
		2.6.6 Effect of Applied Pressure on Permeability	31
		2.6.7 Critical Hydraulic Gradient	32
	2.7	Effective Stresses and Total Stresses	33

	2.8	Consolidation	33
		2.8.1 Computation of Final Consolidation Settlement	34
		2.8.2 Time Rate of Consolidation	35
		2.8.3 Coefficient of Volume Compressibility m_v	39
		2.8.4 Secondary Compression	41
	2.9	Shear Strength	43
		2.9.1 Shear Strength, Friction Angle and Cohesion	43
		2.9.2 Undrained and Drained Loadings in Clays	44
		2.9.3 Undrained Shear Strength of Clays	44
		2.9.4 Peak, Residual and Critical States	45
		2.9.5 Dilatancy Angle	46
		2.9.6 Coefficient of Earth Pressure at Rest	46
	2.10	Soil Variability	47
	References		48
3	**Correlations for Laboratory Test Parameters**		51
	3.1	Permeability	51
		3.1.1 Granular Soils	52
		3.1.2 Cohesive Soils	53
	3.2	Consolidation	54
		3.2.1 Compression Index	54
		3.2.2 Recompression Index or Swelling Index	57
		3.2.3 Compression Ratio and Recompression Ratio	57
		3.2.4 Constrained Modulus	58
		3.2.5 Coefficient of Consolidation c_v	59
		3.2.6 Secondary Compression	60
	3.3	Shear Strength Parameters c' and ϕ'	60
		3.3.1 Cohesion in Terms of Effective Stress c'	60
		3.3.2 Effects of Dilatancy in Granular Soils	62
		3.3.3 ϕ'_{peak}, ϕ'_{cv}, ϕ'_{res} Relationships with Plasticity Index for Clays	63
		3.3.4 Other Friction Angle Correlations	66
		3.3.5 Stress Path Dependence of Friction Angles	69
		3.3.6 Skempton's Pore Pressure Parameters	72
		3.3.7 Sensitivity of Clays	72
	3.4	Undrained Shear Strength of a Clay c_u	73
	3.5	Soil Stiffness and Young's Modulus	76
	3.6	Coefficient of Earth Pressure at Rest K_o	79
	3.7	Using Laboratory Test Data in Pile Designs	81
	References		83
4	**Standard Penetration Test**		87
	4.1	Standard Penetration Test Procedure	87
	4.2	Correction of N Value for Effective Overburden Pressure (For Granular Soils)	89

	4.3	Correction for SPT Hammer Energy Efficiency	91
	4.4	Correlation of Standard Penetration Number with Relative Density (D_r) of Sand	93
	4.5	Correlation of N with Peak Drained Friction Angle (ϕ) for Sand	99
	4.6	Correlation of N with Modulus of Elasticity (E) for Sandy Soils	102
	4.7	Correlation of Undrained Cohesion (c_u) with N for Clay Soil	103
	4.8	Correlation of Preconsolidation Pressure (σ_c') with N for Clay Soil	104
	4.9	Correlation of Overconsolidation Ratio (Ocr) with N for Clay Soil	105
	4.10	Correlation of Cone Penetration Resistance (q_c) with N	105
	4.11	Correlation of Liquefaction Potential of Sand with N	106
	4.12	Correlations for Shear Wave Velocity, v_s	107
	4.13	SPT Correlations with Foundation Bearing Capacity	107
	References		110
5	**Cone Penetrometer Test**		115
	5.1	Cone Penetrometer Test – General	115
	5.2	Piezocone Test – Equipment and Procedure	117
	5.3	Practical Use of Penetrometer Test Results	121
	5.4	Soil Classification	121
	5.5	Correlations for Sands	125
		5.5.1 Correlation with Relative Density of Sand	125
		5.5.2 Correlation of q_c with Sand Friction Angle, ϕ'	128
		5.5.3 Correlation with Constrained Modulus of Cohesionless Soils	130
		5.5.4 Correlation with Small Strain Shear Modulus of Cohesionless Soils	132
	5.6	Correlations for Cohesive Soils	134
		5.6.1 Correlation with Undrained Shear Strength of Cohesive Soils	134
		5.6.2 Correlation with Sensitivity of Cohesive Soils	135
		5.6.3 Correlation with Over Consolidation Ratio of Cohesive Soils	136
		5.6.4 Correlation with Constrained Modulus of Cohesive Soils	137
		5.6.5 Correlation with Compressibility of Cohesive Soils	138
		5.6.6 Correlation with Friction Angle of Cohesive Soils	139
		5.6.7 Correlation with Small Strain Shear Modulus of Cohesive Soils	139
	5.7	Correlation with Unit Weight	141
	5.8	Correlation with Permeability	142

	5.9	Correlation with SPT N	142
	5.10	Correlation with Bearing Capacity	145
		5.10.1 Shallow Foundations	145
		5.10.2 Deep Foundations	146
	5.11	Liquefaction Assessment	148
		5.11.1 Cyclic Stress Ratio	149
		5.11.2 Normalization of Resistance	150
		5.11.3 Computation of Cyclic Resistance Ratio (CRR)	152
	References	153
6	**Pressuremeter Test**	159
	6.1	Pressuremeter Test – General	159
		6.1.1 Menard Type Pressuremeter	160
		6.1.2 Self-Boring Pressuremeter	162
		6.1.3 Other Developments	164
	6.2	Pressuremeter Test – Theoretical Interpretation	165
	6.3	Parameter Derivation	165
		6.3.1 In-Situ Lateral Stress	165
		6.3.2 Young's Modulus	167
		6.3.3 Undrained Shear Strength in Clay	168
		6.3.4 Friction Angle in Sands	171
	6.4	Correlations with Other Tests	173
		6.4.1 Correlation Between Limit Pressure from Menard Type Pressuremeter and q_c from Cone Penetrometer Test	173
		6.4.2 Correlations with Other Soil Parameters – Menard Type Pressuremeter	173
	6.5	Use of Menard Type Pressuremeter Test Results Directly in Design	173
		6.5.1 Ultimate Bearing Capacity (q_u) of Shallow Foundations – Menard Type Pressuremeter	174
		6.5.2 Ultimate Bearing Capacity of Deep Foundations – Menard Type Pressuremeter Test	176
		6.5.3 Skin Friction for Deep Foundations – Menard Type Pressuremeter	176
		6.5.4 Correlation with q_c and SPT N	176
		6.5.5 Other Design Parameters from Menard Type Pressuremeter	177
	References	179
7	**Dilatometer Test**	183
	7.1	Introduction	183
	7.2	Intermediate DMT Parameters	186
	7.3	Correlations	187
	7.4	Summary	191
	References	191

Chapter 1
Introduction

Abstract This chapter is an introduction to the book. It discusses laboratory and in situ tests, their advantages and limitations. The chapter introduces the test covered by the book and discusses the necessity of empirical relationships for the practising engineer. Finally, it briefly mentions how the book is organised into the nine different chapters.

Keywords Laboratory test • In situ test • Empirical correlations

Geotechnical engineering deals with soil and rock, their characteristics and behavior and their effects on design and construction. It covers the broad spectrum of civil engineering including slopes, foundations, embankments and levees, retaining walls, soil nails, anchors, excavations and fills, and the list goes on. As geotechnical engineers our main objective is to understand the behavior of soil and rock, and provide appropriate advice to control and mitigate geotechnical risks associated with any project, large or small. Such advice has to depend on deriving parameters required to assess, analyse and solve problems using the following tools:

1. Laboratory testing
2. In situ testing
3. Trials and/or monitoring during construction

This book covers the first two items listed above.

When discussing geotechnical testing, it is difficult to say which is more appropriate, in situ testing or laboratory testing, as it depends on the particular project and its constraints, as well as objectives of the development.

1.1 Laboratory Testing

Laboratory testing cannot be conducted unless samples are collected from the site which means some form of in situ testing, i.e. at least a borehole or a test pit. Depending on the laboratory test to be conducted, a disturbed, bulk or an undisturbed sample will be required. Most geotechnical investigations use a combination of in situ testing and laboratory testing to assess soil properties. While some may be biased against one or the other, a reasonable geotechnical engineer will assess the objectives of the investigation and the materials encountered or likely to be encountered prior to determining which tests would assist his objective to deliver sound geotechnical advice.

The advantages and limitations for laboratory tests could be summarized as follows (modified from Jamiolkowski et al. 1985):

Advantages – Laboratory Tests

- Well defined boundary conditions.
- Strictly controlled drainage conditions.
- Pre-selected and well defined stress paths are followed during the tests.
- In principle, uniform strain fields.
- Soil nature and physical features are positively identified.
- Multiple tests can be undertaken if sufficient soil sample is available. Useful if confirmation is required when unusual/unexpected results are obtained from a test.

Limitations – Laboratory Tests

- In cohesive soils, the effects of unavoidable sample disturbance in even so-called "high quality" undisturbed samples are sometimes difficult to assess.
- In cohesionless soils, undisturbed sampling is still an unsolved problem in everyday practice.
- The small volume of laboratory specimens (Rowe 1972) cannot incorporate the frequently present macrofabric and inhomogeneities of natural soil deposits. This leads to doubts as to what extent the field behavior of a large soil mass can be successfully modeled by small scale laboratory tests.
- The factors causing the formation of the shear plane during the testing of laboratory specimens are still very poorly understood. It must be emphasized that once a shear plane has developed, deformations are concentrated along this plane and displacements and stresses measured at the specimen boundaries are consequently no longer a function of the stress strain behavior of the tested material.
- In principle, the discontinuous nature of information obtained from laboratory tests may lead to erroneous modeling of the behavior of a large mass.
- In general terms, soil explorations based on the laboratory testing of soil samples obtained from borings are likely to be more expensive and time consuming than explorations which make use of in situ testing techniques.
- For some laboratory tests, operator error could have a significant influence on the end results.

1.2   In Situ Testing

There are many types of in situ testing methods, some more appropriate for cohesive materials and the others more suitable for cohesionless materials. It is not always possible to assess the likely soil materials expected at the site to identify the most suitable test prior to commencing the investigation.

Table 1.1 Applicability and usefulness of in situ tests

Group	Device	Soil parameters													Ground type						
		Soil type	Profile	u	$*\phi'$	c_u	I_D	m_v	c_v	k	G_0	σ_h	OCR	$\sigma-\varepsilon$	Hard Rock	Soft Rock	Gravel	Sand	Silt	Clay	Peat
Penetrometers	Dynamic	C	B	–	–	C	C	–	–	–	–	–	C	–	–	C	B	A	B	B	B
	Mechanical	B	A/B	–	C	C	B	C	–	–	–	–	C	–	–	C	C	A	A	A	A
	Electric (CPT)	B	A	–	C	B	A/B	C	–	–	C	C	C	–	–	C	C	A	A	A	A
	Piezocone (CPTU)	A	A	A	B	B	A/B	B	A/B	B	B	B/C	B	–	–	–	C	A	A	A	A
	Seismic (SCPT/SCPTU)	A	A	A	B	B	A/B	B	A/B	B	A	B/C	B	C	–	–	C	A	A	A	A
	Flat Dilatometer (DMT)	B	A	C	B	B	C	B	–	–	B	B	B	B	–	–	C	A	A	A	A
	Standard Penetration Test (SPT)	A	B	–	C	C	B	–	–	–	C	–	C	–	–	–	C	B	A	A	A	A
Pressuremeters	Resistivity Probe	B	B	–	B	C	A	C	–	–	–	–	–	–	–	–	C	A	A	A	A
	Pre-Bored (PBP)	B	B	–	C	B	C	B	C	–	B	C	C	–	C	–	A	B	B	B	B	A
	Self boring (SBP)	B	B	A[1]	B	B	B	B	A[1]	B	A[2]	A/B	A/B	A/B[2]	A	B	–	B	B	A	A	B
	Full displacement (FDP)	B	B	–	C	B	C	C	C	–	A[2]	C	C	C	–	–	C	B	B	B	A	A
Others	Vane	B	C	–	–	A	–	–	–	–	–	–	B/C	B	–	–	–	–	–	–	A	B
	Plate load	C	–	–	C	B	B	B	C	–	C	–	B	B	B	A	B	B	A	A	A
	Screw plate	C	C	–	C	B	B	B	C	–	C	–	B	–	–	–	–	A	A	A	A
	Borehole permeability	C	–	A	–	–	–	–	B	A	–	–	–	–	A	A	A	A	A	A	B
	Hydraulic fracture	–	–	B	–	–	–	–	–	–	B	–	–	–	B	–	A	–	–	–	C	–
	Crosshole/downhole/surface seismic	C	C	–	–	–	–	–	–	–	A	–	B	–	A	A	A	A	A	A	A

Applicability: A = High, B = Moderate, C = Low, – = None
[1] = only when pore pressure sensor fitted, [2] = only when displacement sensor fitted *ϕ' = will depend on soil type

Soil parameter definition: u = in situ pore pressure; ϕ' = effective internal friction angle, c_u = undrained shear strength, m_v = coefficient of volume compressibility
c_v = coefficient of consolidation, k = coefficient of permeability, G_0 = shear modulus at small strains; OCR = over consolidation ratio σ_h = horizontal stress
$\sigma - \varepsilon$ = stress–strain relationship; I_D = density index

As presented in Table 1.1, Lunne et al. (1997) summarized the various in situ tests in operation at the time and classified them according to their applicability and usefulness in deriving different design parameters as well as in different material types. There have been advances since the publication of this table, especially in piezocone and dilatometer testing, which are covered later in this book.

The advantages and limitations for in situ tests could be summarized as follows (modified from Jamiolkowski et al. 1985):

Advantages – In Situ Tests

- A larger volume of soil is tested than is usually done in most laboratory tests; hence in situ tests should in principle reflect more accurately the influence of macrofabric on the measured soil characteristics.
- Several in situ tests produce a continuous record of the soil properties which allows the soil macrofabric and layer boundaries to be determined.
- In situ tests can be carried out in soil deposits in which undisturbed sampling is still impossible or unreliable. Examples include cohesionless soils, soils with highly-developed macrofabrics, intensively layered and/or heterogeneous soils, and highly fissured clays.
- The soils are tested in their natural environment which may not be preserved in laboratory tests. For example, the most successful attempts to measure the existing initial total in situ horizontal stress are the recent developments in situ techniques, e.g. self boring pressuremeter, flat dilatometer, Iowa stepped blade, spade-like total stress cells.
- Some in situ tests are relatively inexpensive compared to investigations based on laboratory tests.
- In general terms, soil exploration by means of in situ techniques is less time consuming than investigations based on laboratory tests.
- Results from in situ tests are readily available and could be interpreted with minimum delay compared to the delivery of sample to a laboratory and time for the test to be carried out. This is most significant when the test site is a long distance from a laboratory.

Limitations – In Situ Tests

- Boundary conditions in terms of stresses and/or strains are, with the possible exception of the self boring pressuremeter, poorly defined, and a rational interpretation of in situ tests is very difficult.
- Drainage conditions during the tests are generally unknown and make it uncertain if the derived soil characteristics reflect undrained, drained or partially drained behavior. In this respect, quasistatic cone penetration tests with pore pressure measurements and self boring pressuremeter tests (also with pore pressure measurements), when properly programmed, help to minimize the problem.
- The degree of disturbance caused by advancing the device in the ground and its influence on the test results is generally (with the possible exception of the self boring pressuremeter) large but of unknown magnitude.

- Modes of deformation and failure imposed on the surrounding soil are generally different from those of civil engineering structures; furthermore, they are frequently not well established, as for example in the field vane test.
- The strain fields are nonuniform and strain rates are higher than those applied in the laboratory tests or those which are anticipated in the foundation on structure.
- With the exception of the Standard Penetration test which allows the collection of samples, the nature of the tested soil is not directly identified by in situ tests.

In this book we have concentrated on only a select few in situ tests which we believe are used more frequently than others. They include the following:

- Standard Penetration test,
- Cone Penetrometer test,
- Pressuremeter test,
- Dilatometer test and
- Vane Shear test.

Summary of advantages and limitations relevant to these tests are presented in Table 1.2 (modified from Becker, 2001).

1.3 Empirical Correlations

While in situ testing and laboratory testing on samples recovered during a site investigation remain the two preferred methods for determining the design parameters in geotechnical engineering, there is a substantial cost associated with these two methods. In the preliminary stages including feasibility studies, when there is limited funds available for soil exploration, empirical correlations become very valuable. For example, from simple index properties, one can get a fair idea about the shear strength and consolidation characteristics of a clay at little or no cost.

In geotechnical engineering, empiricism has a big role to play. In addition to giving preliminary estimates, the correlations can also be used to compare against the values determined from laboratory and in situ tests. There are so many empirical equations and graphs available in the literature, which are regularly being used in the designs worldwide. These are derived based on laboratory or field data, past experience, and good judgement.

In US, the Army, Navy and Air Force do excellent engineering work, and invest significantly in research and development. All their design guides, empirical equations, and charts are very well proven and tested. They are generally conservative, which is not a bad thing in engineering. Most of these manuals are available for free download. They (e.g. NAVFAC 7.1) are valuable additions to your professional library.

The Canadian Foundation Engineering Manual is a well respected design manual used in Canada. Kulhawy and Mayne (1990) produced a good report that takes a close look at the different empirical correlations and charts, in the light of more

Table 1.2 Summary of advantages and limitations of in situ tests covered by this book

Test	Advantages	Limitations
Standard penetration test (SPT)	• Standardized test that is robust, inexpensive • Feasible to carry out in a wide range of materials • Provides a sample (split spoon) • Widely used for many years and has a large database and correlations for most engineering properties • Basis of design for foundations and liquefaction assessment of materials	• Affected by borehole disturbance, such as piping, base heave and stress relief • Affected by equipment to make borehole, energy efficiency and by operator • Results influenced by grain size, soil structure and stress history • Many corrections required for interpretation and design
Piezocone penetration test (CPTu)	• Robust and easy to use standardized test • Continuous profiles obtained • Relatively quick test and a large number of tests can be done in a day • No need for a borehole unless obstructions encountered. • If carried out properly, test results are accurate and repeatable • Different measurements made, which enhance interpretation • Increasingly used in liquefaction assessment of materials • Many correlations available for most engineering properties and design applications • Avoids disturbance effects associated with boreholes	• Not suitable for materials with large particles, which obstruct penetration. Best suited for uniform, fine grained soils • Not easy to penetrate very dense or hard materials • Problems can develop with rod buckling when hard material is suddenly encountered under softer soils • Needs calibration against other tests to obtain strength and stiffness data • No sample is obtained • Instrument relatively expensive
• Pressuremeter test and self boring pressuremeter	• The stress strain curve can be derived; not just a single value of an engineering property • Boundary conditions are controlled and well defined. Testing carried out at both small and large strains • Self boring pressuremeter can be inserted in suitable soils with minimal disturbance and avoiding stress relief effects • Useful for determination of in situ horizontal stress • Use of loading and unloading cycles can mitigate borehole and other disturbances effects and enhance interpretation • Correlations available for	• No sample is obtained and test results should be supported by other strength data • Sophisticated, relatively expensive instrument requiring experienced, skilled operators • Testing is time consuming and on a less continuous basis than other tests such as CPTu • Test results affected by procedure, and method of interpretation is important. Different methods of interpretation give different results • Borehole required for Menard type pressuremeter and some soils • Effects of disturbance and stress

(continued)

1.3 Empirical Correlations

Table 1.2 (continued)

Test	Advantages	Limitations
	important engineering properties and design applications	relief need to be considered and taken into account
Dilatometer test (DMT)	• Robust, simple standardized test, easy to carry out using same equipment as other tests • Testing on a near-continuous basis • Results generally repeatable • Avoids disturbance effects of boreholes • Correlations available for important engineering properties and design applications • Good test for interpretation of in situ horizontal stress and deformation at small strain	• Not suited for soils with large particles or that are too dense or hard or hard to permit penetration without use of borehole • No sample obtained • Different methods of interpretation give different results • Testing more time consuming and less continuous basis than tests such as CPTu • Limited strain imposed during test. Results not suited for large strain behavior
Vane shear test (VST)	• Robust, simple standardized test that is easy to carry out • Direct measurement of shear strength • Only in situ test that provides direct measurement of residual strength • In some cases, no borehole is needed; but rod friction needs to be eliminated or measured	• Generally limited to clays which shear strength <150 kPa • Not suitable for most other soils • Results affected by sandy/silty layers and gravel inclusions • Generally corrections required • Results may be misleading in some soils (e.g. silts) • Gives shear strength in the horizontal direction • May not give operational strength in fissured soils • Not suitable for materials with large particles, which obstruct penetration; Best suited for uniform, fine grained soils • Difficult to penetrate very dense or hard materials • Problems can develop with rod buckling when a hard material is suddenly encountered under softer soils

data. They also came up with their own correlations and simplifications, which are quite popular among practicing engineers.

Geotechnical data, whether from the field or laboratory, can be quite expensive. We often have access to very limited field data (e.g. SPT) from a few boreholes, along with some laboratory test data on samples obtained from these bore holes and/or trial pits. We use the empirical correlations sensibly to extract the maximum possible information from the limited laboratory and field data which come at a high price.

1.4 Contents of the Book

The main objective of this book is to provide readers correlations commonly used by geotechnical practitioners to assess design parameters important in the geotechnical design activities. Such correlations may be the sole weapon available to the designer when no or poor soil or rock data is available and/or only limited testing has been carried out. In some situations, it is simply not practical to carry out complex or expensive testing, whether in situ or laboratory, because the gains are not significant and the risks or opportunities are very little. In other instances, correlations are required as a test of reasonableness on the derived design parameters. Generally, at concept design stage of a project, most practitioners are left with the use of correlations and typical values as site investigation data may not be available or limited.

Chapter 2 provides an overview of geotechnical properties commonly used in the designs and analysis, definitions of main soil mechanics terms and phase relationships.

In Chapter 3, laboratory tests required to obtain geotechnical properties of soils, and the empirical correlations relating to the different parameters are discussed.

Chapter 4 is reserved for the Standard Penetration test, popularly known as SPT, a simple test but that provides very useful data on the resistance of soils which could be translated into strength and stiffness as well as to obtain many other geotechnical parameters.

Chapter 5 describes the Cone Penetrometer test (CPT) which is increasingly used by practitioners because the test is easy to carry out and quick. It has become a routine test for site investigations worldwide to characterize clays and sands.

In Chapter 6, the Pressuremeter is described and its use to derive geotechnical design parameters using standard correlations is explored. The test is unique amongst in situ tests because it can be performed in soft clays to weak rock.

Chapter 7 describes the Dilatometer test, the test measurements and the methods for determining the design parameters and the soil types. Chapter 8 covers the vane shear test, most popular for the testing of soft to firm clayey soils.

Chapter 9, the last chapter, is devoted to rocks, and includes some understanding of their properties, the different tests carried out on intact rock specimens and the correlations that can be used to estimate the properties of the intact rock and the parent rock mass.

References

Becker DE (2001) Chapter 4: Site characterization. In: Rowe RK (ed) Geotechnical and geoenvironmental engineering handbook. Kluwer Academic Publishers, Norwell, pp 69–105

Jamiolkowski M, Ladd CC, Germaine JT, Lancellotta R (1985) New developments in field and laboratory testing of soils. In: Proceedings 11th international conference on soil mechanics and foundation engineering, vol 1, San Francisco, pp 57–154

Kulhawy FH, Mayne PW (1990) Manual on estimating soil properties for foundation design. Report EL- 6800 submitted to Electric Power Research Institute, Palo Alto, California, 306 p

Lunne T, Robertson PK, Powell JJM (1997) Cone penetration testing in geotechnical practice. Blackie Academic/Chapman-Hall Publishers, London, 312p

Rowe PW (1972) The relevance of soil fabric to site investigation practice. 12th Rankine Lecture, Geotechnique, No 2

Chapter 2
Geotechnical Properties of Soils – Fundamentals

Abstract This chapter gives a brief overview of the geotechnical properties commonly determined in the laboratory, their relevance in soil mechanics and laboratory tests for determining them. The properties discussed include Atterberg limits, the different densities, particle size distribution, permeability, and the parameters related to consolidation and shear strength. The tests required to obtain these parameters are also discussed. The information in this chapter gives the necessary background to understand the empirical correlations relating the different parameters determined in the laboratory.

Keywords Soil properties • Design parameters • Soil mechanics • Permeability • Consolidation • Shear strength

This chapter gives a brief overview of the geotechnical properties commonly determined in the laboratory, their relevance in soil mechanics and laboratory tests for determining them. It gives the necessary background for Chap. 3, which covers the empirical correlations relating the different parameters determined in the laboratory. For an in-depth understanding of the soil mechanics principles, the reader is referred to the geotechnical engineering textbooks. The laboratory test procedures are covered in good detail in some specialized references such as Das (2009), Sivakugan et al. (2011), etc.

2.1 Laboratory Tests for Soils

Laboratory soil tests are carried out on intact or disturbed soil samples collected from the site. In granular (cohesionless) soils, it is very difficult to obtain intact samples and therefore, their soil parameters are determined indirectly through in situ (field) tests. Alternatively, laboratory tests can be carried out on reconstituted granular soils, where the grains are packed at densities that match the in situ soils. In cohesive soils, intact samples are obtained generally in sampling tubes which are capped, the ends waxed and wrapped in Polythene bags to maintain the moisture until they are extruded and tested in the laboratory (Fig. 2.1). The sampling tubes

Fig. 2.1 Soil samples from the site showing the sampling date, project and borehole numbers, and the depth

are clearly labeled, showing the project number, borehole number, date of sampling, and sample depth, as shown in Fig. 2.1.

Laboratory tests and in situ tests complement each other. One can never be the substitute of the other. They have their advantages and disadvantages, and hence a well-designed site investigation program with a good balance of laboratory and in situ tests can be very effective in deriving the design parameters. Laboratory tests are carried out under well-defined boundary conditions, on small specimens that are often homogeneous. This makes the laboratory tests reproducible, and the interpretation of the laboratory test data is generally carried out using rational soil mechanics principles. The small specimen size and the effort involved in testing the specimen, makes it difficult to test large volumes.

2.2 Phase Relations

The soil consists of three phases: soil grains (i.e., solids), water, and air. Their relative proportions are represented schematically as shown in Fig. 2.2 in a *phase diagram*, where the volumes are shown on the left and the masses are shown on the right, denoted by V and M, respectively. The subscripts s, w, a, and v denote soil

Fig. 2.2 Phase diagram

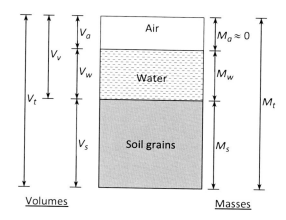

grains, water, air and void, respectively. V_t and M_t represent the total volume and total mass, which includes the soil grains, water and air.

2.2.1 Terminology and Definitions

The *water content* w, generally expressed as a percentage, is defined as:

$$w(\%) = \frac{M_w}{M_s} \times 100 \tag{2.1}$$

It is also known as *moisture content*. Dry soils have $w = 0\,\%$. In organic or soft clays and peats, water content can exceed 100 %. The water content of a soil in its in situ natural state is known as the *natural water content* (w_n). The soils near the surface are rarely dry and they absorb some water from the atmosphere, and remain at *hygroscopic water content*.

Void ratio (e) and *porosity* (n) are both measures of the void volume within the soil. They are defined as:

$$e = \frac{V_v}{V_s} \tag{2.2}$$

and

$$n(\%) = \frac{V_v}{V_t} \times 100 \tag{2.3}$$

Void ratio can be as low as 0.3 in compacted or well graded soils (significantly less in rocks) to very much greater than 2.0 for soft clays or organic soils such as peats. Theoretically, porosity lies in the range of 0–100 %.

Degree of saturation S defines the percentage of the void volume that is filled by water, and is defined as:

$$S(\%) = \frac{V_w}{V_v} \times 100 \qquad (2.4)$$

It varies between 0 % for dry soils and 100 % for saturated soils. The *air content* a is a term that is used with compacted soils. It is a measure of the air volume, expressed as a percentage of the *total volume*, and hence defined as:

$$a(\%) = \frac{V_a}{V_t} \times 100 \qquad (2.5)$$

Density ρ is the mass per unit volume. There are few different ways of defining the density, which include bulk density, dry density, saturated density and submerged density. The *bulk density* ρ_m is the ratio of M_t to V_t, where the soil can contain all three phases. It is also known as total, moist or wet density. When the soil is dried at the same void ratio, with air in the entire voids, the *dry density* ρ_d is the ratio of M_s to V_t. When the soil is saturated at the same void ratio, the voids are filled with water and the soil has only two phases, and the bulk density is known as *saturated density* ρ_{sat}. Submerged density ρ' accounts for the buoyancy effects under water, and is defined as:

$$\rho' = \rho_{sat} - \rho_w \qquad (2.6)$$

where ρ_w is the density of water, which is 1.0 g/cm³, 1.0 t/m³, 1.0 Mg/m³ or 1000 kg/m³. Density becomes the unit weight (γ), when the mass is replaced by the weight. When using phase relations, it is possible to work with densities (and masses) or unit weights (and weights), as long as proper units are maintained. Note that $\gamma = \rho g$ where g is the gravity (9.81 m/s²). Bulk, dry, saturated and submerged unit weights are denoted by $\gamma_m, \gamma_d, \gamma_{sat}$, and γ', respectively. The unit weight of water γ_w is 9.81 kN/m³. The unit weights of saturated soils vary in the range of 15–22 kN/m³, with cohesive soils at the lower end of the range and granular soils at the upper end. Remember that the unit weight of concrete is about 24 kN/m³, which can be seen as an upper limit for soils. Rocks generally have unit weights greater than that of concrete.

The water content of a wet soil can be determined in the laboratory, by noting the wet mass of a sample and the dry mass after drying it in an oven at 105 °C for 24 h (ASTM D2216; BS 1377-2; AS 1289.2.1.1). Knowing the volume of the specimen, it is possible to determine the density or unit weight.

Specific gravity (G_s) of the soil grains is often required in the computations of the masses and volumes of the three phases. Specific gravity is a measure of how heavy the soil grains are compared to water. It is defined as

$$G_s = \frac{\text{Density of soil grains}}{\text{Density of water}} \qquad (2.7)$$

Determining of specific gravity, using a *density bottle* or *pycnometer*, is fairly straightforward exercise that uses Archimedes' principle (ASTM D854; BS 1377-2; AS 1289.3.5.1).

Table 2.1 Typical values of G_s

Mineral	G_s	Mineral	G_s
Quartz[a]	2.65	K-Feldspar[a]	2.55
Na-Ca Feldspar[a]	2.6–2.8	Calcite[a]	2.72
Dolomite[a]	2.85	Muscovite[a]	2.7–3.1
Biotite[a]	2.8–3.2	Chlorite[a]	2.6–2.9
Pyrophyllite[a]	2.84	Serpentine[a]	2.2–2.7
Kaolinite[a]	2.64	Halloysite (2H$_2$O)[a]	2.55
Illite[a]	2.84	Montmorillonite[a]	2.76
Attapulgite[a]	2.30	Loess from central US[b]	2.70
Volcanic ash, Kansas[b]	2.32	Micaceous silt, Alaska[b]	2.76
Gypsum[c]	2.3–2.4	Galena[c]	7.4–7.6
Pyrite[c]	4.9–5.2	Magnetite[c]	4.4–5.2

Adapted from Lambe and Whitman (1979), Handy and Spangler (2007), and Winchell (1942)
[a]Lambe and Whitman (1979)
[b]Handy and Spangler (2007)
[c]Winchell (1942)

For most soils G_s varies in a narrow range of 2.6–2.8. For mine tailings rich in heavy minerals, G_s can be as high as 4.0 or even larger, and for lighter materials such as fly ash, peat and organic soils, it can be significantly less than the lower end of the above range. Typical values of G_s for different minerals are summarized in Table 2.1.

2.2.2 Relationships Between the Variables

Phase relations are the equations that relate the masses and volumes of the three different phases. There are several different parameters (e.g., e, n, S, w, G_s, ρ_d, etc.) that were defined in this section. One can write dozens of equations expressing one variable in terms of few others. The three major phase relations that are adequate for computing the masses and volumes of the different phases are given below [Eqs. (2.8), (2.9), and (2.10)]. Proving these relations is a fairly straightforward exercise that is discussed in most geotechnical textbooks (Das 2010; Holtz et al. 2011; Sivakugan and Das 2010). In these equations, w, n and S are expressed as decimal numbers instead of percentages.

$$w = \frac{Se}{G_s} \quad (2.8)$$

$$n = \frac{e}{1+e} \quad (2.9)$$

$$\rho_m = \left(\frac{G_s + Se}{1+e}\right)\rho_w \quad (2.10)$$

Substituting $S = 0$ or 1 in Eq. (2.10), the expressions for dry and saturated densities can be deduced as

$$\rho_d = \left(\frac{G_s}{1+e}\right)\rho_w \qquad (2.11)$$

$$\rho_{sat} = \left(\frac{G_s+e}{1+e}\right)\rho_w \qquad (2.12)$$

Dry density is related to bulk density by

$$\rho_d = \frac{\rho_m}{1+w} \qquad (2.13)$$

which is a useful relationship in compaction, for computing ρ_d from ρ_m. The air content a [see Eq. (2.5)] is a term used in compaction that can be expressed as

$$a = \frac{e(1-S)}{1+e} \qquad (2.14)$$

The Eqs. (2.10), (2.11), (2.12), and (2.13) apply for unit weights as well, where ρ is replaced by γ on both sides of the equations.

A good knowledge of typical values for unit weights of different soils is required for estimating the overburden stresses at different depths. The unit weight can vary in the range of 15–21 kN/m^3 for most soils, depending on whether they are saturated or not. Some typical values suggested by the Australian standard for earth retaining structures (AS 4678-2002) are given in Table 2.2.

It can be seen from Table 2.2 that there is no difference in the unit weight of the bulk and saturated weights of cohesive soils. In granular soils, the difference is slightly greater for loose material than dense ones.

2.3 Granular Soils

Grain size distribution quantifies the relative proportions of the different grain sizes present within a soil. Relative density is a measure of how densely (or loosely) the grains are packed within a specific grain size distribution. These two terms are discussed below.

2.3.1 Grain Size Distribution

Grain size distribution plays a major role in how the granular soils behave. This is not the case with the clays, where the mineralogy and Atterberg limits become more important. The grain size distribution test is generally carried out using sieves (ASTM D6913; BS1377-2; AS 1289.3.6.1) and hydrometer (ASTM D422;

2.3 Granular Soils

Table 2.2 Typical values for bulk and saturated unit weights

		Bulk unit weight (kN/m^3)		Saturated unit weight (kN/m^3)	
		Loose	Dense	Loose	Dense
Granular soils	Gravel	16.0	18.0	20.0	21.0
	Well graded sand and gravel	19.0	21.0	21.5	23.0
	Coarse or medium sand	16.5	18.5	20.0	21.5
	Well graded sand	18.0	21.0	20.5	22.5
	Fine or silty sand	17.0	19.0	20.0	21.5
	Rock fill	15.0	17.5	19.5	21.0
	Brick hardcore	13.0	17.5	16.5	19.0
	Slag fill	12.0	15.0	18.0	20.0
	Ash fill	6.5	10.0	13.0	15.0
Cohesive soils	Peat (high variability)	12.0		12.0	
	Organic clay	15.0		15.0	
	Soft clay	17.0		17.0	
	Firm clay	18.0		18.0	
	Stiff clay	19.0		19.0	
	Hard clay	20.0		20.0	
	Stiff or hard glacial clay	21.0		21.0	

After AS 4678-2002

BS1377-2; AS 1289.3.6.3). A set of sieves is used for separating the different sizes of coarse grained (gravels and sands) soils and a hydrometer is used on the fine grained (silts and clays) soils. The resulting grain size data are plotted as the percentage passing versus grain size (logarithmic scale). D_{10}, D_{15}, D_{30}, D_{50}, D_{60}, D_{85} are some of the common grain sizes derived from the *grain size distribution curve* for soil classification, designs of filters and vibroflotation. D_{10} is the grain size corresponding to 10 % passing. In other words, 10 % of the grains are smaller than this size. D_{10} is also known as the *effective grain size*, which reflects the size of the pore channels that conduct water through soils. D_{50} is the *median grain size* – 50 % of the grains are larger than this size, which is different from the *mean* grain size. The spread of the grain sizes within the soil is reflected in the magnitude of the *coefficient of uniformity* (C_u) defined as:

$$C_u = \frac{D_{60}}{D_{10}} \qquad (2.15)$$

C_u is always greater than unity. A value close to unity implies that the grains are about the same size. Another grain size distribution parameter that is used in classifying granular soils is the *coefficient of curvature* or *coefficient of gradation* C_c defined as:

$$C_c = \frac{D_{30}^2}{D_{10}D_{60}} \qquad (2.16)$$

Sands are considered *well graded* when $C_u > 6$ and $C_c = 1\text{--}3$. Gravels are well graded when $C_u > 4$ and $C_c = 1\text{--}3$. Well graded soils have a wide range of grain sizes present. Coarse grained soils that do not meet these criteria are classified as poorly graded soils, which includes uniformly graded soils and gap graded soils. Gap graded soils are the ones where there are little or no grains in a specific size range. Coarse grained soils are generally classified on the basis of the grain size distribution, and fine grained soils based on Atterberg limits which take into account the soil plasticity.

2.3.2 Relative Density

A granular soil can be packed to different densities. Its strength and stiffness are determined by the state of packing. The *maximum void ratio* and the *minimum dry density*, which occur at the loosest possible state, are denoted by e_{max} and $\rho_{d,min}$, respectively. The *minimum void ratio* and the *maximum dry density* which take place at the densest possible state are denoted by e_{min} and $\rho_{d,max}$, respectively. They are easily determined by laboratory tests (ASTM D4253/4254; BS1377-4; AS 1289.5.5.1).

The relative packing of the grains within a granular soil is quantified through *relative density* D_r (also known as *density index* I_d) defined as:

$$D_r(\%) = \frac{e_{max} - e}{e_{max} - e_{min}} \times 100 \qquad (2.17)$$

where, e is the void ratio of the current state at which the relative density is being determined. Between the loosest and the densest state, D_r varies from 0–100 %. Granular soils can be classified on the basis of D_r as suggested by Lambe and Whitman (1979) and shown in Fig. 2.3. The term relative density should not be used in granular soils containing more than 15 % fines. In terms of densities, D_r can be written as:

$$D_r(\%) = \frac{\rho_{d,max}}{\rho_d} \times \frac{\rho_d - \rho_{d,min}}{\rho_{d,max} - \rho_{d,min}} \times 100 \qquad (2.18)$$

where, ρ_d is the dry density at which the relative density is being determined. Equation (2.18) can also be written in terms of unit weights.

2.4 Plasticity

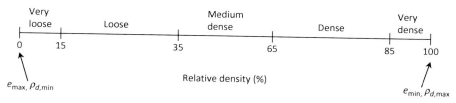

Fig. 2.3 Classification of granular soils based on relative density D_r

Table 2.3 Typical values for e_{max} and e_{min}

Soil	e_{max}	e_{min}	References
Uniform sub-angular sand	0.85	0.50	Sowers and Sowers (1961)
Well-graded sub-angular sand	0.70	0.35	
Very well graded silty sandy gravel	0.65	0.25	
Micaceous sand and silt	1.25	0.80	
Well graded fine to coarse sand	0.70	0.35	McCarthy (1977)
Uniform fine to medium sand	0.85	0.50	
Silty sand and gravel	0.80	0.25	
Micaceous sand and silt	1.25	0.80	
Uniform spheres	0.92	0.35	Lambe and Whitman (1979)
Standard Ottawa sand	0.80	0.50	
Clean uniform sand	1.00	0.40	
Uniform inorganic silt	1.10	0.40	
Silty sand	0.90	0.30	
Fine to coarse sand	0.95	0.20	
Micaceous sand	1.20	0.40	
Silty sand and gravel	0.85	0.14	

Some typical values for e_{max} and e_{min} are given in Table 2.3. It can be seen that for granular soils e_{min} can be in the range of 0.2–0.7 and e_{max} can be in the range of 0.7–1.2.

2.4 Plasticity

Plasticity is a term that is associated with clays. The mineralogy of the clay grains, their grain shapes resembling flakes and needles with large surface area per unit mass, and the charge imbalance make them cohesive and plastic. Gravels, sands and silts are non-plastic.

2.4.1 Atterberg Limits

Liquid limit (LL or w_L), *plastic limit* (PL or w_P), and *shrinkage limit* (SL or w_S) are known as the *Atterberg limits* which define the borderline water contents that

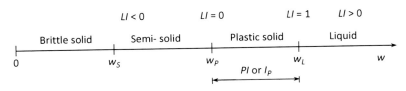

Fig. 2.4 Atterberg limits – The borderline water contents

separate the different states at which a *fine grained* soil can exist, as shown in Fig. 2.4. The soil remains plastic when the water content lies between *LL* and *PL*, and the difference between *LL* and *PL* is known as the *plasticity index* (*PI* or I_P). When dried below shrinkage limit, there will be no volume reduction (i.e., shrinkage) with the loss of moisture. The soil becomes unsaturated, losing some water in the voids, while there is no further volume reduction. Liquid limit is determined in the laboratory by Casagrande's percussion cup method (ASTM D4318; BS 1377-2; AS 1289.3.1.1) or Swedish fall cone method (BS1377-2; AS 1289.3.9.1). Plastic limit is defined as the lowest water content at which the fine grained soil can be rolled into a 3 mm diameter thread (ASTM D4318; BS1377-2; AS 1289.3.2.1).

Liquidity index (*LI*) is defined as:

$$LI = \frac{w - PL}{LL - PL} \quad (2.19)$$

where *w* is the water content at which *LI* is being determined. *LI* is a measure of where the current water content (*w*) lies with respect to the *PL-LL* range. In bore logs, it is a common practice to show the *LL-PL* range and the natural water content with respect to this range for clay soils at different depths. The natural water content lying to the right of this range implies *LI* greater than 1. Left of this range, *LI* is negative. *Consistency index* (*CI*) is a term used sometimes in the literature, which is defined as:

$$CI = 1 - LI = \frac{LL - w}{LL - PL} \quad (2.20)$$

Plasticity is associated with clays. Plasticity index is a measure of plasticity. Pure silts are non-plastic, with $PI \approx 0$. Burmister (1949) classified cohesive soils on the basis of plasticity, as shown in Table 2.4. Table 2.4 suggests *PI* as the single measure of plasticity. It may not necessarily be the case. As Casagrande (1948) noted, *PI* should be seen along with *LL* when defining plasticity.

2.4.2 Classification of Fine Grained Soils Based on Plasticity

PI and *LL* are the main parameters used in the classification of fine grained soils including silty clays and clayey silts. The Unified Soil Classification System

Table 2.4 Classification of clays based on *PI*

PI	Classification
0	Non-plastic
1–5	Slightly plastic
5–10	Low plastic
10–20	Medium plastic
20–40	High plastic
>40	Very high plastic

(USCS) is the most widely used soil classification system. The fine grained soils are classified and assigned a two-letter symbol based on *LL* and *PI* as shown in Casagrande's (1948) *PI-LL* chart in Fig. 2.5. The first letter denotes the soil group (i.e. silt, clay or organic soil) and the second letter describes the plasticity.

A soil that lies above the A-line is a clay (symbol C) and below the line it is classified as silt (symbol M). U-line is the upper limit below which all soils are expected to lie. Depending on whether *LL* is less or greater than 50, the fine grained soil is classified as a soil with low (L) or high (H) plasticity. The fine grained soil can have a symbol of CL, CH, ML or MH. When there is significant organic content in the soil, it can be classified as organic soil, and assigned symbol of OH or OL depending on the liquid limit. Peats have symbol of Pt. When the values of *LL* and *PI* are such that the soil lies within the shaded region in Fig. 2.5, the fine grained soil is classified as silty clay or clayey silt, with symbol of CL-ML.

Plastic limit and liquid limit tests are generally carried out on the soil fraction passing No. 40 (0.425 mm) sieve. This fraction can contain clays, silts and some fine sands. Two clays having the same plasticity index can have quite different behavior depending on their mineralogical characteristics and the clay content. They can be distinguished by *activity*, denoted by *A* and defined as

$$\text{Activity} = \frac{\text{Plasticity index}}{\text{Percentage of clay fraction}} \quad (2.21)$$

This is a good measure of the potential swell problems in clays. Typical values of activity for different clay minerals are shown in Table 2.5. Clays with activity larger than one may show very high swell potential. Such clays are known as *expansive clays* or *reactive clays*. Due to repeated wetting and drying, they undergo swell-shrink cycles and cause significant damage to the infrastructure such as light buildings and roads.

2.5 Compaction

Compaction is the simplest form of ground improvement that is carried out to improve the ground conditions, prior to the construction of buildings, roads, embankments, etc. The effectiveness of the compaction depends on the type of

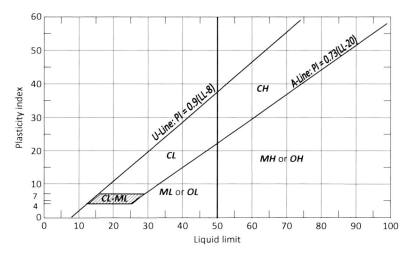

Fig. 2.5 Casagrande's *PI-LL* chart for classification of fine grained soils

Table 2.5 Typical values of activity

Mineral	Activity
Calcite	0.2
Kaolinite	0.3–0.5
Illite	0.5–1.3
Ca- Montmorillonite	1.5
Na- Montmorillonite	4–7
Muscovite	0.2
Quartz	0

After Skempton (1953) and Mitchell (1976)

rollers used, number of passes, and the moulding water content. Typical variation of the dry density of the compacted soil against the water content is shown in Fig. 2.6. At the optimum water content (w_{opt}) the dry density is the maximum ($\rho_{d,\max}$) and the void ratio is the minimum. Compacted soils show excellent engineering properties at or near the optimum water content. Nevertheless, in the case of clay soils, there can be significant variation in the strength, stiffness, permeability, swell/shrink potential and the fabric depending on the water content. Clays compacted dry of optimum are brittle, stronger, stiffer, and will have greater swell potential and are more permeable. Depending on the nature of the compacted earthwork, the compaction is carried out slightly dry or wet of optimum.

In any soil, the optimum water content and the maximum dry density vary with the *compactive effort*. Higher the compactive effort, higher the maximum dry density and lower the optimum water content. To effectively specify and control the field compaction, it is necessary to know w_{opt} and $\rho_{d,max}$ of the soil under the specific compactive effort. These are two of the key parameters that are used as the basis for specifying the compaction requirements. They are generally determined

2.5 Compaction

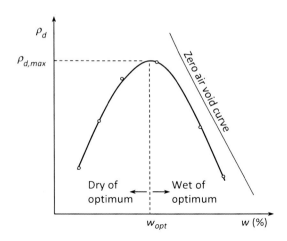

Fig. 2.6 Moisture-density relation in compaction

through standard (ASTM D698; BS1377-4; AS 1289.5.1.1) or modified (ASTM D1557; BS1377-4; AS 1289.5.2.1) Proctor compaction tests in the laboratory. The *zero air void curve* in Fig. 2.6 is the locus of $S = 100\ \%$. All compaction test points should lie to the left of the zero air void curve, implying $S < 100\ \%$. A point lying to the right of the zero air void curve suggests $S > 100\ \%$, which is not possible. The zero air void curve will shift upwards with larger G_s values. Therefore, it is necessary to use the right value for G_s when plotting the zero air void curve.

Standard Proctor and modified Proctor are the two different compactive efforts adopted commonly in earthworks. Standard Proctor compactive effort applies 592 kN·m/m^3 and the modified Proctor compactive effort applies 2678 kN·m/m^3.

For standard Proctor compactive effort on fine grained soils, the optimum water content is about 2–4 % less than the plastic limit, with high plastic clays falling at the lower end of the range. This can be used as a guide in estimating the optimum water content without actually doing the compaction test. In sands, the lowest water content at which a drop of water can be squeezed out is approximately the optimum water content. Fine grained soils, compacted to standard Proctor compactive effort, are at 80–85 % degree of saturation at the optimum water content. When they are compacted wet of optimum, the degree of saturation is in the range of 90–95 % where the compaction curve is roughly parallel to the zero air void curve.

The differences in the optimum water content and maximum dry density, between the standard and modified Proctor compaction tests, are more pronounced for clays than sands. The maximum dry density under a standard Proctor compactive effort can be about 85–97 % of the modified Proctor compactive effort. The optimum water content under modified Proctor compactive effort can be 2–5 % lower than that of standard Proctor compactive effort (Hausmann 1990).

These days, modified Proctor compactive effort is specified more commonly, and some typical requirements as suggested by U.S. Navy (1982) and Hausmann (1990) are summarized in Table 2.6. Relative desirability ratings of different types of soils for specific applications are summarized in Table 2.7.

Table 2.6 Typical compaction requirements

Fill used for	% of $\gamma_{d,\,max}$ from modified Proctor	Water content range about optimum (%)
Roads:		
Depth of 0–0.5 m	90–105[a]	−2 to +2
Depth > 0.5 m	90–95[a]	−2 to +2
Small earth dam	90–95	−1 to +3
Large earth dam	95	−1 to +2
Railway embankment	95	−2 to +2
Foundation for structure	95	−2 to +2
Backfill behind walls/trenches	90	−2 to +2
Canal linings of clays	90	−2 to +2
Drainage blanket or filter	90	Thoroughly wet

After US Navy (1982) and Hausmann (1990)
[a]Depending on soil type, traffic and function of the layer

2.6 Permeability

Soils are *porous media* which allow water to flow through the interconnected voids. Permeability k is a measure of how easily the water can flow through the soils.

2.6.1 D'Arcy's Law and Permeability Measurements

In laminar flow through soils, the velocity v is proportional to the *hydraulic gradient i*. They are related by (D'Arcy 1856)

$$v = ki \tag{2.22}$$

The hydraulic gradient is the *total* head loss per unit length along the flow path, and is a dimensionless number. The permeability, also known as *hydraulic conductivity*, has the unit of velocity. It is commonly expressed in cm/s or m/s.

Permeability of the intact or reconstituted specimens can be measured in the laboratory by *constant head* or *falling head permeability tests*, that are suited for coarse grained soils (ASTM D2434; BS1377-6; AS 1289.6.7.1) and fine grained soils (BS1377-6; AS 1289.6.7.2), respectively. In the field, pump-in or pump-out tests are carried out within wells for computing the permeability. Here, water is pumped into or out of the wells until steady state is reached, when the water levels at the observation wells are measured and the permeability is computed. Simple field infiltrometers can be used for estimating the average saturated in situ permeability of the ground. A metal ring is driven a short distance into the ground and filled with water. The wetting front advances downward and the hydraulic gradient

2.6 Permeability

Table 2.7 Relative desirability of soils as compacted fills

Relative desirability for various uses (No. 1 is considered the best; No. 14 is the least desirable; "—" = not suitable)

USCS Symbol	Soil type	Rolled earth fill dams			Canal sections		Foundations		Roadways		
		Homogeneous embankment	Core	Shell	Erosion resistance	Compacted earth lining	Seepage important	Seepage not important	Fills		Surfacing
									Frost heave not possible	Frost heave possible	
GW	Well graded gravels, gravel-sand mixtures, little or no fines	—	—	1	1	—	—	1	1	1	3
GP	Poorly graded gravels, gravel sand mixtures, little or no fines	—	—	2	2	—	—	3	3	3	—
GM	Silty gravels, poorly graded gravel-sand-silt mixtures	2	4	—	4	4	1	4	4	9	5
GC	Clayey gravels, poorly graded gravel-sand-clay mixtures	1	1	—	3	1	2	6	5	5	1
SW	Well graded sands, gravelly sands, little or no fines	—	—	3 if gravelly	6	—	—	2	2	2	4
SP	Poorly graded sands, gravelly sands, little or no fines	—	—	4 if gravelly	7 if gravelly	—	—	5	6	4	—
SM	Silty sands, poorly graded sand-silt mixtures	4	5	—	8 if gravelly	5 erosion critical	3	7	6	10	6
SC	Clayey sands, poorly graded sand-clay mixtures	3	2	—	5	2	4	8	7	6	2
ML	Inorganic silts and very fine sands, rock flour, silty or clayey fine sands with slight plasticity	6	6	—	—	6 erosion critical	6	9	10	11	—
CL	Inorganic clays of low to medium plasticity, gravelly	5	3	—	9	3	5	10	9	7	7

(continued)

Table 2.7 (continued)

USCS Symbol	Soil type	Relative desirability for various uses (No. 1 is considered the best; No. 14 is the least desirable; "–" = not suitable)								
		Rolled earth fill dams			Canal sections			Foundations		Roadways	
										Fills	
		Homogeneous embankment	Core	Shell	Erosion resistance	Compacted earth lining		Seepage important	Seepage not important	Frost heave not possible	Frost heave possible	Surfacing
OL	clays, sandy clays, silty clays, lean clays											
	Organic clays organic silt-clays of low plasticity	8	8	–	–	7 erosion critical		7	11	11	12	–
MH	Inorganic silts, micaceous or diatomaceous fine sandy or silty soils, elastic silts	9	9	–	–	–		8	12	12	13	–
CH	Inorganic clays of high plasticity, fat clays	7	7	–	10	8 volume change critical		9	13	13	8	–
OH	Organic clays of medium-high plasticity	10	10	–	–	–		10	14	14	14	–

U.S. Navy (1982)

2.6 Permeability

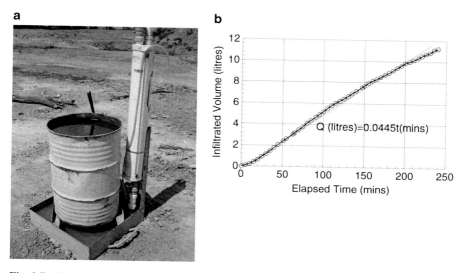

Fig. 2.7 Field infiltrometer: (**a**) Field setup, and (**b**) Test data

can be assumed as unity, which is reasonable in granular soils. In clayey soils, matric suction comes into play, and makes this assumption questionable. A double-ring infiltrometer, with water filling the space between the two rings, can ensure that the flow beneath the inner ring is vertical.

Figure 2.7a shows a square-base infiltrometer with 750 mm × 750 mm base dimensions, and a flow meter to measure the flow rate. The drum filled with water simply provides the vertical load required to hold the infiltrometer firmly seated into the soil. At steady state, the flow rate can be calculated from the data shown in Fig. 2.7b as 0.0445 l/min. Assuming hydraulic gradient of unity and applying D'Arcy's law, permeability can be estimated as 1.3×10^{-4} cm/s.

Permeability has slight dependence on the temperature. Higher the temperature lower the viscosity and higher the permeability. The standard practice is to report the value at 20 °C (ASTM D2434). It can be obtained from:

$$k_{20°C} = k_{T°C} \frac{\mu_{T°C}}{\mu_{20°C}} \tag{2.23}$$

where $\mu_{20°C}$ = dynamic viscosity at 20 °C, and $\mu_{T°C}$ = dynamic viscosity at $T°C$. ASTM D5084 suggests that $\mu_{T°C}/\mu_{20°C}$ can be approximated as:

$$\frac{\mu_{T°C}}{\mu_{20°C}} = \frac{2.2902 \times 0.9842^T}{T^{0.1702}} \tag{2.24}$$

Noting that the flow takes place only through a fraction of the cross section, the actual velocity of the water (or the pore fluid), known as *seepage velocity* v_s, is higher than the discharge velocity v [Eq. (2.22)] generally used in soil mechanics. It can be determined as:

$$v_s = \frac{v}{n} \qquad (2.25)$$

where, n is the porosity, expressed as a decimal number. In geotechnical engineering computations, when applying D'Arcy's law, the discharge velocity is used. However, the seepage velocity is the real velocity of a water molecule.

Permeability of a soil depends on the grain size distribution, void ratio, the soil fabric and the degree of saturation. The permeability of an unsaturated soil is often significantly less than that of the saturated one. When dealing with groundwater and seepage problems, full saturation is often assumed. The values and the empirical correlations for permeability discussed in this chapter are those for *saturated* soils.

2.6.2 Intrinsic Permeability

The permeability k, as defined in Eq. (2.22) depends on the hydraulic properties (e.g., viscosity and density) of the permeant. Within the same porous soil skeleton, the flow characteristics can be quite different for water and oil. In petroleum industry, geologists and engineers deal with flow of oil through rocks. *Intrinsic permeability* or *absolute permeability* K is introduced to eliminate this dependence of permeability on the hydraulic properties, defining K as

$$K = \frac{\mu_w}{\gamma_w} k \qquad (2.26)$$

where μ_w and γ_w are the dynamic viscosity and the unit weight of water, respectively. The dimension of K is L^2 with unit of cm^2, m^2, etc. This intrinsic permeability of the soil matrix K is a measure of the geometry and size of the void network, and is independent of the permeant characteristics. In oil and gas industry, Darcy is a common unit for K (1 Darcy $= 0.987$ μm^2). For flow of water through soils, assuming $\gamma_w = 9810$ N/m^3 and $\mu_w = 1.002 \times 10^{-3}$ N·s/m^2 at 20 °C,

$$K(\text{cm}^2) = k(\text{cm/s}) \times 1.02 \times 10^{-5} \qquad (2.27)$$

$$K(\text{Darcy}) = k(\text{cm/s}) \times 1.035 \times 10^3 \qquad (2.28)$$

Kenney et al. (1984) carried out laboratory permeability studies on compacted granular soils and concluded that:

$$K(\text{mm}^2) = (0.5 \text{ to } 0.1) D_5^2 \qquad (2.29)$$

where D_5 is in mm.

2.6 Permeability

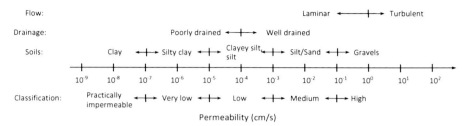

Fig. 2.8 Typical values of permeability

2.6.3 Reynold's Number and Laminar Flow

Figure 2.8 can be used as a rough guide for the flow (laminar or turbulent) and drainage (well or poorly drained) characteristics of the major soil groups, and the ranges of their permeability values. A simple classification based on permeability as suggested by Terzaghi and Peck (1967) is also shown. Reynold's number R for the flow through soils can be defined as

$$R = \frac{vD\rho_w}{\mu_w} \qquad (2.30)$$

where v = discharge velocity, D = average diameter of the pores, ρ_w = density of water (1000 kg/m^3), and μ_w = dynamic viscosity of water (1.002× 10^{-3} N·s/m^2) which is also known as the *absolute viscosity*. The ratio μ_w/ρ_w is the kinematic viscosity of water. Muskat (1946) and Scheidegger (1957) discussed the early experimental work that has been carried by several researchers out to determine the limiting value of R beyond which the flow will not be laminar and D'Arcy's law becomes invalid. It appears that $R = 1$ can be taken as a conservative estimate. Flow through coarse sands and gravels is generally turbulent. In computing Reynolds number from Eq. (2.30), D is sometimes taken as D_{10} or the average grain diameter.

2.6.4 Anisotropy

The permeability of cohesive soils can be *anisotropic*, where it is generally larger in the horizontal direction than in the vertical direction. In special case of loess deposits, the vertical permeability can be larger than the horizontal permeability (Harr 1962). The ratio k_h/k_v reported in the literature is generally less than 2 for most soils. Fukushima and Ishii (1986) showed that for a weathered granite, compacted at different water contents, this ratio to be quite high, sometimes exceeding 10. In varved clays and stratified fluvial deposits, this ratio can easily exceed 10 (Casagrande and Poulos 1969; Tavenas and Leroueil 1987; Wu et al. 1978). Some anisotropic behavior of natural clays reported by Tavenas and

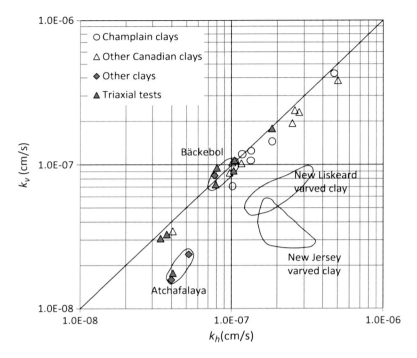

Fig. 2.9 Permeability anisotropy in natural clays (Adapted from Tavenas and Leroueil 1987)

Leroueil (1987) are shown in Fig. 2.9. Varved clays are naturally deposited layered soils of different grain sizes that occur due to seasonal fluctuations of sediment conditions in glacial lakes. Figure 2.10 shows a photograph of varved clays from Connecticut, USA.

The flow nets for solving seepage problems are generally drawn such that the stream lines and the equipotential lines intersect at 90°. This is true only when the soil is isotropic. In anisotropic soils, where the horizontal and vertical permeabilities k_h and k_v are different, they do not intersect at 90°, which makes it difficult to draw the flow net. The common practice is to transform the anisotropic soil system into an equivalent isotropic soil system where the dimensions along the horizontal direction are scaled (i.e., shrunk) by $\sqrt{\frac{k_v}{k_h}}$. The isotropic permeability of the equivalent soil system is taken as $\sqrt{k_h k_v}$.

2.6.5 One-Dimensional Flow in Layered Soils

Let's consider a layered soil system where the layer thicknesses are H_1, H_2, ...H_i,...H_n, and the corresponding permeabilities of the layers are k_1, k_2...k_i,...k_n.

Fig. 2.10 Varved clay (Courtesy of Natural Resources Conservation Services, US Department of Agriculture)

Within the layers, the permeability is isotropic. If there is one dimensional flow taking place parallel to the layers, the layered system can be analyzed as an equivalent homogeneous soil layer with thickness of $H_1 + H_2 + \ldots + H_n$. The equivalent permeability of this layer is given by

$$k_{eq//} = \frac{k_1 H_1 + k_2 H_2 + \ldots + k_n H_n}{H_1 + H_2 + \ldots + H_n} \qquad (2.31)$$

If one dimensional flow takes place perpendicular to the layers, the equivalent layered system can be analyzed as an equivalent homogeneous layer with thickness of $H_1 + H_2 + \ldots + H_n$. The equivalent permeability of this layer can be obtained from Eq. (2.32).

$$\frac{H_1 + H_2 + \ldots + H_n}{K_{eq\perp}} = \frac{H_1}{k_1} + \frac{H_2}{k_2} + \ldots + \frac{H_n}{k_n} \qquad (2.32)$$

2.6.6 *Effect of Applied Pressure on Permeability*

Most of the time constant head or falling head permeability tests are carried out in the laboratory under no surcharge or applied pressure. In reality, there can be

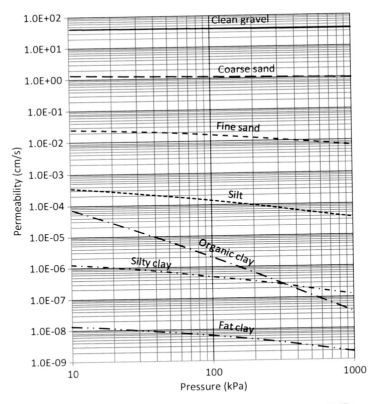

Fig. 2.11 Variation of permeability with surcharge (Adapted from Cedegren 1967)

significant effective stress acting on the soil due to the overburden and the loads applied at the ground level. Therefore, it is useful to know the effect of the applied surcharge on permeability. Cedegren (1967) suggested the trends and values shown in Fig. 2.11.

2.6.7 *Critical Hydraulic Gradient*

In flow through soils, the seepage force per unit volume of the soil is expressed as $i\gamma_w$. During upward flow in granular soils, the hydraulic gradient can become high enough to induce a seepage force that exceeds the submerged weight of the soil grain, thus causing failure within the soil mass through a mechanism known as *piping* or *heave*. This situation occurs when the hydraulic gradient (i) exceeds the *critical hydraulic gradient* (i_c) given by γ'/γ_w. It can be shown from simple phase relations that:

$$i_c = \frac{\gamma'}{\gamma_w} = \frac{G_s - 1}{1 + e} \qquad (2.33)$$

With typical value of $G_s \approx 2.65$ for sands, and void ratios in the range of 0.5–0.8, the critical hydraulic gradient is generally about unity.

2.7 Effective Stresses and Total Stresses

In a saturated soil, the *total normal stress* σ at a point is shared by the soil grains and the pore water. The component carried by the soil grains (i.e., the soil skeleton) is known as the *effective stress* or *intergranular stress* σ' and the component carried by the water is called *pore water pressure* or *neutral stress* u. The total and effective stresses are related by:

$$\sigma = \sigma' + u \qquad (2.34)$$

Pore water pressure is hydrostatic, meaning, it is the same in all directions. The effective and total stresses are directional. Equation 2.34 is valid at all times at any point within the soil mass, and is applicable in any direction. Water cannot carry any shear stress. Therefore, the entire shear stress is carried by the soil skeleton.

Pore water pressure can be negative. For example, when there is capillary effect, the soil is subjected to suction and the pore water pressure is negative. Dynamic loads such as earthquakes or pile driving can increase the pore water pressures temporarily, while the total stress remains the same. Here, the effective stress is temporarily reduced [see Eq. (2.34)]. Such dynamic loads can induce *liquefaction* of the soil.

2.8 Consolidation

Consolidation is a time-dependant mechanical process where some of the water in a saturated soil is squeezed out of the voids by the application of external loads. It occurs almost instantaneously in coarse grained soils, but takes a long time in cohesive soils due to their low permeability. In general, consolidation theory is applied to *saturated* cohesive soils, where the process takes from several weeks to several years, depending on the consolidation characteristics (i.e. compressibility of the soil skeleton, and permeability) of the clay, thickness of the clay layer, and the drainage conditions at the boundaries.

Let's consider a saturated clay layer of thickness H shown in Fig. 2.12a. The initial void ratio of the clay is e_0, and the clay is at its natural water content w_n and current effective vertical overburden stress of σ'_{vo} at the middle of the clay layer. When the clay is loaded or surcharged at the ground level, the void ratio decreases

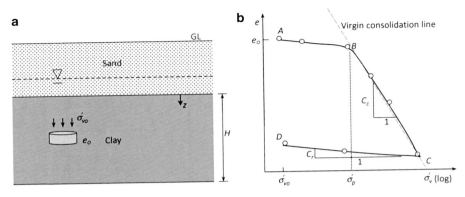

Fig. 2.12 Consolidation: (**a**) Soil profile, and (**b**) $e - \log \sigma'_v$ plot

and the effective stress increases along the path *ABC* shown in Fig. 2.12b. The sudden break at *B* defines the *preconsolidation pressure* σ'_p, which is the maximum pressure the in situ clay has experienced ever in its past geological history. Beyond the preconsolidation pressure, the $e - \log \sigma'_v$ variation is approximately linear (sector *BC*) with a slope of C_c, known as the *compression index*. When the surcharge is completely removed, the clay layer heaves with the void ratio increasing, and the $e - \log \sigma'_v$ variation is again linear (sector *CD*), with a slope of C_s, known as the *swelling index* (or *recompression index* C_r). When reloaded, approximately the same unloading path *DC* is followed. C_c and C_r are measures of how compressible the saturated soil skeleton is.

2.8.1 Computation of Final Consolidation Settlement

When the clay layer is subjected to some loading at the ground level, consolidation takes place. One of the main objectives in the consolidation analysis is to estimate the *final consolidation settlement* (s_c) that will occur at the end of the consolidation process, under a specific load applied at the ground level. This can be done in two ways: Using the coefficient of volume compressibility (m_v), or by using the change in void ratio (Δe). These are discussed next.

(a) Determining s_c using m_v:

The *coefficient of volume compressibility* m_v, is defined as the ratio of volumetric strain to the applied effective stress. In one-dimensional consolidation, m_v can be written as:

$$m_v = \frac{1}{D} = \frac{\Delta H / H}{\Delta \sigma'} \qquad (2.35)$$

where ΔH is the change in thickness (same as s_c) during the consolidation under the effective normal stress increase of $\Delta \sigma'$, and *H* is the initial thickness at the beginning of consolidation. The reciprocal of m_v is known as the *constrained modulus* or *oedometer modulus D*, which is a measure of stiffness, when the soil is restrained from any lateral deformation. It is similar to the Young's modulus E determined without any lateral constraints. The problem with m_v is that unlike C_c and C_r, it varies with the stress level. It is necessary to know the appropriate value of m_v at the relevant stress level for estimating the final consolidation settlement realistically. The final consolidation settlement can be estimated from

$$s_c = \Delta H = m_v \Delta \sigma' H \tag{2.36}$$

where $\Delta \sigma'$ is the normal stress increase at the middle of the clay layer, and H is the thickness of the clay layer. The unit of m_v is kPa^{-1} or MPa^{-1}.

(b) Determining s_c using change in void ratio Δe:

Using the initial void ratio e_0, compression index C_c, recompression index C_r, preconsolidation pressure σ'_p, initial effective vertical stress σ'_{vo}, and the increase in vertical stress $\Delta \sigma'$, the reduction in the void ratio Δe due to complete consolidation can be computed. The final consolidation settlement s_c can be computed from

$$s_c = \frac{\Delta e}{1 + e_0} H \tag{2.37}$$

2.8.2 Time Rate of Consolidation

Terzaghi (1925) showed that the three variables excess pore water pressure u, the depth within the clay z, and the time t since application of the load, are related by the following governing differential equation:

$$\frac{\partial u}{\partial t} = c_v \frac{\partial^2 u}{\partial z^2} \tag{2.38}$$

where c_v is the *coefficient of consolidation*, defined as

$$c_v = \frac{k}{m_v \gamma_w} \tag{2.39}$$

where γ_w is the unit weight of water ($=9.81$ kN/m^3) and k is the permeability of the clay.

Consolidation test is commonly carried out on an intact clay specimen in a rigid metal ring known as *oedometer* (ASTM D2435; AS 1289.6.6.1) or in a *Rowe cell* (BS1377-5) which is common in Europe. The load is applied in increments, with 24 h between the increments, allowing for full consolidation under each increment. From the consolidation test data, the $e - \log \sigma'_v$ plot (see Fig. 2.12b) can be generated and the parameters such as C_c, C_r, σ'_p, m_v, and c_v can be determined.

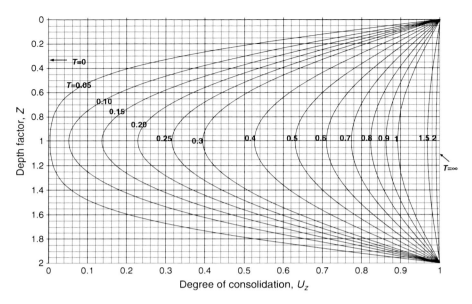

Fig. 2.13 U-Z-T relationship

The Eq. 2.36 or 2.37 only gives the final consolidation settlement that occurs at the end of the consolidation process. The fraction of the excess pore water pressure that has dissipated at a specific depth z (see Fig. 2.12a) at a specific time t is called the *degree of consolidation* $U(z,t)$, which varies with depth and time. It is given by

$$U(z,t) = 1 - \sum_{m=0}^{m=\infty} \frac{2}{M} \sin(MZ) e^{-M^2 T} \qquad (2.40)$$

where $M = (\pi/2)(2m+1)$. Z and T are the dimensionless *depth factor* and *time factor* defined as

$$Z = \frac{z}{H_{dr}} \qquad (2.41)$$

and

$$T = \frac{c_v t}{H_{dr}^2} \qquad (2.42)$$

Here, H_{dr} is the maximum length of the drainage path, which is H for singly drained layers and $H/2$ for doubly drained layers, where H is the thickness of the clay layer (see Fig. 2.12a). The interrelationship among U, Z and T is shown in Fig. 2.13. This figure can be used to determine the excess or undissipated pore water pressure at any depth at any time.

2.8 Consolidation

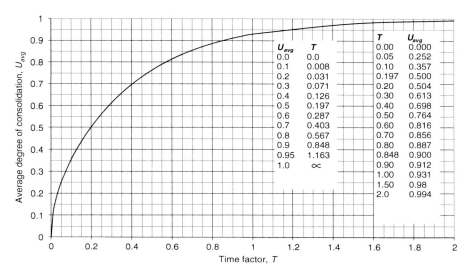

Fig. 2.14 $U_{avg} - T$ relationship

The average degree of consolidation $U_{avg}(t)$ of the clay layer at time t, is given by

$$U_{avg} = 1 - \sum_{m=0}^{m=\infty} \frac{2}{M^2} e^{-M^2 T} \qquad (2.43)$$

U_{avg} is the same as the fraction of the consolidation settlement that has taken place at time t. U_{avg}-T relation is shown in Fig. 2.14. This can be approximated as (Terzaghi 1943)

$$T = \frac{\pi}{4} U_{avg}^2 \qquad \text{for } U_{avg} \leq 52.6\% \qquad (2.44)$$

$$T = 1.781 - 0.933 \log(100 - U_{avg}) \qquad \text{for } U_{avg} \geq 52.6\% \qquad (2.45)$$

The U_{avg}-T chart in Fig. 2.14 is developed assuming that the initial excess pore water pressure that drives the consolidation process is uniform or linearly increases or decreases with depth, and the clay layer is doubly drained. The same chart also applies when the initial excess pore water pressure is uniform and singly drained. These are adequate for most practical applications. The U_{avg}-T values for other initial excess pore water pressure distributions, for singly and doubly drained situations, are given in Table 2.8.

Constant Rate Loading

In carrying out the consolidation analysis and computing the settlements, it is implied that the entire load is applied instantaneously. In reality, this is never the

Table 2.8 U_{avg}-T values for different initial pore water pressure distributions for singly and doubly drained clay layers

	U_{avg} (Singly drained – impervious base)					U_{avg} (Doubly drained)		
T	Uniform	Lin Inc	Lin Dec	Half-sine	Sine	Uniform/linear	Half-sine	Sine
0.004	0.0714	0.0080	0.1374	0.0098	0.0194	0.0714	0.0584	0.0098
0.008	0.1009	0.0160	0.1859	0.0195	0.0380	0.1009	0.0839	0.0195
0.012	0.1236	0.0240	0.2232	0.0292	0.0558	0.1236	0.1040	0.0292
0.020	0.1596	0.0400	0.2792	0.0482	0.0896	0.1596	0.1366	0.0482
0.028	0.1888	0.0560	0.3216	0.0668	0.1207	0.1888	0.1637	0.0668
0.036	0.2141	0.0720	0.3562	0.0850	0.1495	0.2141	0.1877	0.0850
0.048	0.2472	0.0960	0.3985	0.1117	0.1887	0.2472	0.2196	0.1117
0.060	0.2764	0.1199	0.4329	0.1376	0.2238	0.2764	0.2481	0.1376
0.072	0.3028	0.1436	0.4620	0.1628	0.2553	0.3028	0.2743	0.1628
0.083	0.3251	0.1651	0.4851	0.1852	0.2816	0.3251	0.2967	0.1852
0.100	0.3568	0.1977	0.5159	0.2187	0.3184	0.3568	0.3288	0.2187
0.125	0.3989	0.2442	0.5536	0.2654	0.3659	0.3989	0.3719	0.2654
0.150	0.4370	0.2886	0.5853	0.3093	0.4077	0.4370	0.4112	0.3093
0.167	0.4610	0.3174	0.6045	0.3377	0.4337	0.4610	0.4361	0.3377
0.175	0.4718	0.3306	0.6130	0.3507	0.4453	0.4718	0.4473	0.3507
0.200	0.5041	0.3704	0.6378	0.3895	0.4798	0.5041	0.4809	0.3895
0.250	0.5622	0.4432	0.6813	0.4604	0.5413	0.5622	0.5417	0.4604
0.300	0.6132	0.5078	0.7187	0.5230	0.5949	0.6132	0.5950	0.5230
0.350	0.6582	0.5649	0.7515	0.5784	0.6420	0.6582	0.6421	0.5784
0.400	0.6979	0.6154	0.7804	0.6273	0.6836	0.6979	0.6836	0.6273
0.500	0.7640	0.6995	0.8284	0.7088	0.7528	0.7640	0.7528	0.7088
0.600	0.8156	0.7652	0.8660	0.7725	0.8069	0.8159	0.8069	0.7725
0.700	0.8559	0.8165	0.8953	0.8222	0.8491	0.8559	0.8491	0.8222
0.800	0.8874	0.8566	0.9182	0.8611	0.8821	0.8874	0.8821	0.8611
0.900	0.9120	0.8880	0.9361	0.8915	0.9079	0.9120	0.9079	0.8915
1.000	0.9313	0.9125	0.9500	0.9152	0.9280	0.9313	0.9280	0.9152
1.500	0.9800	0.9745	0.9855	0.9753	0.9790	0.9800	0.9790	0.9753
2.000	0.9942	0.9926	0.9958	0.9928	0.9939	0.9942	0.9939	0.9928

Courtesy of Dr. Julie Lovisa, James Cook University

case. Generally, a building or embankment is constructed in stages, and the construction can take several months. Assuming constant rate of loading, where the load is applied over a period t_0 at a constant rate, the average degree of consolidation U_{avg} at any time $(t < t_0)$ during the construction can be written as (Sivakugan and Vigneswaran 1991):

$$U_{avg} = 1 - \frac{1}{T}\left[\sum_{m=0}^{m=\infty}\left(\frac{2}{M^4}\right)\left(1 - e^{-M^2 T}\right)\right] \quad (2.46)$$

A plot of U_{avg} against T for constant rate loading is shown in Fig. 2.15. The $U_{\text{avg}} - T$ values are also shown within the figure. For a specific value of T, the U_{avg}

2.8 Consolidation

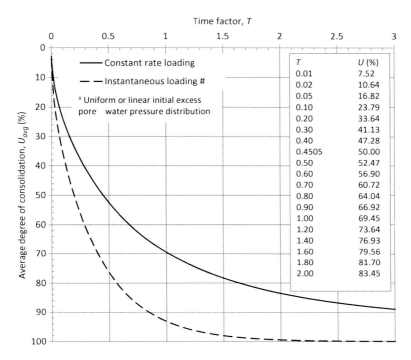

Fig. 2.15 $U_{avg} - T$ relationship for constant rate loading

computed from Eq. (2.46) is always significantly less than that computed from Eq. (2.43). This can also be seen in Fig. 2.15.

It is widely known that the consolidation process in the field is significantly faster than what is expected from the c_v determined in the laboratory. This is probably due to the different boundary conditions, and the sand seams and fissures present in the field that are not present in the small laboratory specimens. Balasubramaniam et al. (2010) reported that the field c_v is 5–10 times larger than the laboratory determined c_v.

2.8.3 Coefficient of Volume Compressibility m_v

The coefficient of volume compressibility m_v typically varies in the range of 0.01–2.0 MPa^{-1}. Larger the m_v softer is the soil skeleton. Table 2.9 shows a simple classification of clays based on m_v.

Some value ranges for different soils, as suggested by Domenico and Mifflin (1965) are shown in Table 2.10.

The coefficient of volume compressibility m_v is the reciprocal of the constrained modulus D. They are related to the Young's modulus by

Table 2.9 Classification of clays based on m_v

Type of soil	m_v (MPa^{-1})	Compressibility
Heavily overconsolidated clays	<0.05	Very low
Very stiff or hard clays, tills	0.05–0.10	Low
Varved and laminated clays, firm to stiff clays	0.10–0.30	Medium
Normally consolidated alluvial clays	0.3–1.5	High
Organic alluvial clays and peats	>1.5	Very high

After Bell (2000)

Table 2.10 Ranges of m_v values for different soils

Soil type	m_v (MPa^{-1})
Plastic clay	0.26 to 2.1
Stiff clay	0.13 to 0.26
Medium hard clay	0.069 to 0.13
Loose sand	0.052 to 0.1
Dense sand	0.013 to 0.021
Dense sandy gravel	$(0.1 \text{ to } 5.2) \times 10^{-3}$
Jointed rock	$(0.33 \text{ to } 6.9) \times 10^{-3}$
Sound rock	$\geq 0.33 \times 10^{-3}$
Water	0.44×10^{-3}

After Domenico and Mifflin (1965)

$$D = \frac{1}{m_v} = \frac{1-\nu}{(1+\nu)(1-2\nu)} E = K + \frac{4}{3}G \qquad (2.47)$$

where, ν = Poisson's ratio, E = Young's modulus, K = bulk modulus, and G = shear modulus. From D or m_v determined from the oedometer, assuming a value for the Poisson's ratio ν, the Young's modulus E can be estimated. For $\nu = 0.1$–0.33, $D = 1.0$–$1.5\,E$. E and ν are related to K and G by

$$K = \frac{E}{3(1-2\nu)} \qquad (2.48)$$

$$G = \frac{E}{2(1+\nu)} \qquad (2.49)$$

In numerical modelling work, K and G are sometimes used as the elastic input parameters than E and ν. E and ν can be expressed in terms of K and G as

$$E = \frac{9KG}{3K+G} \qquad (2.50)$$

$$\nu = \frac{3K-2G}{3(3K+G)} \qquad (2.51)$$

The Young's modulus of clays derived from in situ tests is often under undrained conditions (i.e. E_u) when there is little or no drainage from the soil during the test.

2.8 Consolidation

By equating the shear modulus under drained and undrained conditions using Eq. (2.49),

$$\frac{E_u}{2(1+\nu_u)} = \frac{E}{2(1+\nu)} \qquad (2.52)$$

where E and ν are the drained values. For all saturated clays under undrained loading $\nu_u = 0.5$. Substituting $\nu_u = 0.5$,

$$E = \frac{2}{3}(1+\nu)E_u \qquad (2.53)$$

It is evident from Eq. (2.53) that the drained modulus is slightly less than the undrained modulus. With $\nu_u = 0.5$, and ν of 0.12–0.35, E_u/E lies in the range of 1.11–1.34.

2.8.4 Secondary Compression

According to Terzaghi's one-dimensional consolidation theory, the consolidation is a never ending process that continues indefinitely. In reality, the consolidation process ends after some time when the excess pore water pressure induced by the applied load has fully dissipated. Once the consolidation, also known as *primary consolidation*, is completed, there will still be some continuous reduction in the void ratio and hence an increase in the settlements. This process is known as the *secondary compression* or *creep*, which takes place at constant effective stress when there is no further dissipation of the excess pore water pressure. This occurs due to the reorientation of the clay particles, and other mechanisms which are not properly understood. For simplicity, it is generally assumed that secondary compression begins on the completion of primary consolidation. During the secondary compression, the variation of void ratio with logarithmic time is approximately linear, which enables us to define the *coefficient of secondary compression* (C_α) as:

$$C_\alpha = \frac{\Delta e}{\Delta (\log t)} \qquad (2.54)$$

where Δe and $\Delta(\log t)$ are the changes in void ratio and logarithm of time within a time period during the secondary compression. C_α can be determined from the consolidation test data, or estimated from empirical correlations discussed in Chap. 3.

The secondary compression settlement s_s of a clay layer at time t can be computed by

$$s_s = C_\alpha \frac{H_p}{1+e_p} \log\left(\frac{t}{t_p}\right) \qquad (2.55)$$

where, t_p = time for completion of primary consolidation, H_p = clay layer thickness at the end of primary consolidation, and e_p = void ratio at the end of primary consolidation. Sometimes it can be difficult to get realistic estimates of H_p and e_p. On the other hand H_0 and e_0 at the beginning of consolidation are readily available. Therefore $H_p/(1+e_p)$ in Eq. (2.55) can be replaced by $H_0/(1+e_0)$. The secondary compression settlement s_s computed in Eq. (2.55) is the settlement that takes place between the times t_p and t.

When expressed in terms of vertical strain, instead of void ratio, C_α can be written as

$$C_{\alpha\varepsilon} = \frac{\Delta\varepsilon}{\Delta(\log t)} \qquad (2.56)$$

where, $\Delta\varepsilon$ = vertical strain during the time interval Δt. $C_{\alpha\varepsilon}$ is known as the *modified secondary compression index*. C_α and $C_{\alpha\varepsilon}$ are related by

$$C_{\alpha\varepsilon} = \frac{C_\alpha}{1+e_0} \qquad (2.57)$$

where, e_0 is the void ratio at the beginning of the time interval over which C_α is computed. For normally consolidated clays, $C_{\alpha\varepsilon}$ lies in the range of 0.005–0.02. For highly plastic clays or organic clays, it can be 0.03 or higher. For overconsolidated clays with $OCR > 2$, $C_{\alpha\varepsilon}$ is less than 0.001.

Alonso et al. (2000) suggested that the ratio of C_α for overconsolidated and normally consolidated clays can be written as

$$\frac{C_{\alpha\varepsilon,\ OC}}{C_{\alpha\varepsilon,\ NC}} = (1-m)e^{-(OCR-1)n} + m \qquad (2.58)$$

where, m and n are constants. The constant m is the minimum possible value for the above ratio which applies for very large OCR, and is similar to the C_r/C_c ratio. The magnitude of n controls the rate of decay in the ratio with OCR, with larger values of n giving a faster decay. Alonso et al. (2000) suggested $m = 0.1$ and $n = 12$ from limited data. In practice, smaller values for n are being used conservatively. Wong (2006) suggested $n = 6$ in organic clays for preliminary assessments.

A proper consolidation test, with 6–8 load increments and an unloading cycle, with all the associated measurements, can take about 2 weeks, and can cost $1000 or more. The following parameters can be derived from these tests:

- $e - \log \sigma'_v$ plot which defines C_c and C_r'
- σ'_p based on Casagrande's (1936) graphical procedure
- In situ virgin consolidation line based on Schmertmann's (1955) procedure

- m_v as a function of σ'_v
- c_v as a function of σ'_v
- k as a function of σ'_v
- C_α as a function of σ'_v

2.9 Shear Strength

Stability of an embankment, foundation, or a retaining wall is governed by the shear strength characteristics of the surrounding soil. The salient features of shear strength are briefly discussed in this section.

2.9.1 Shear Strength, Friction Angle and Cohesion

Soils generally fail in shear where the soil grains slide over each other along the failure surface, and not by crushing of soil grains. *Shear strength* τ_f of soils can be described by *Mohr-Coulomb failure criterion* which relates the shear stress τ_f on the failure plane (i.e., shear strength) with the normal stress σ on the same plane and two soil constants: *friction angle* ϕ, and *cohesion c*. According to Mohr-Coulomb failure criterion,

$$\tau_f = \sigma \tan \phi + c \tag{2.59}$$

which is an equation of a straight line shown in Fig. 2.16 where tan ϕ is the slope or gradient of the line and c is the intercept on τ-axis.

It can be seen from Eq. (2.59) that the shear strength of a soil consists of two independent components, which are derived from friction ($\sigma \tan\phi$) and cohesion (c). The frictional component is proportional to the normal stress on the plane, and the cohesive component is independent of the normal stress. For example, when the

Fig. 2.16 Mohr-Coulomb failure envelope

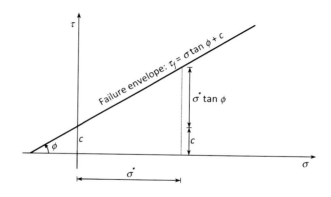

normal stress on the failure plane is σ^*, the shear stress required to cause failure (i.e., the shear strength) is given by $\sigma^* \tan \phi + c$ (see Fig. 2.16). The constants c and ϕ remain the same at all stress levels.

When analyzing in terms of total stresses, the normal stress, friction angle and cohesion are simply denoted by σ, c and ϕ, respectively. In terms of effective stresses, they are denoted by σ', c' and ϕ', respectively. Shear stress is the same in terms of effective and total stresses; pore water does not carry any shear stress.

2.9.2 Undrained and Drained Loadings in Clays

When a building or an embankment load is applied on a clayey ground, it is often assumed that the entire load is applied instantaneously. Soon after the loading (i.e., *short-term*), with very little time for any drainage, it is fair to assume that the clay is loaded under *undrained conditions*. Under undrained conditions, the clay is often analyzed in terms of total stresses, without worrying about the effective stresses and pore water pressures induced within the clay which are often unknown. Here, no attempt is made to separate the stresses carried by the soil skeleton formed by the clay particles and the pore water; the entire clay is treated as one homogeneous material.

After a long time (i.e., *long-term*), it is a different story. The clay would have fully drained, with no *excess* pore water pressures, and the loading can be treated as *drained loading*, where the analysis can be carried out in terms of effective stresses. There can still be static pore water pressures due to the phreatic surface (i.e., water table). Similarly, if it is known that the clay is loaded very slowly without any build-up of excess pore water pressures, such situations can also be analyzed as drained loading, in terms of effective stresses. In granular soils, where the drainage is always good with little or no build-up of excess pore water pressures, all loadings are under drained conditions.

Total stress analysis is carried out in terms of undrained shear strength parameters c_u and ϕ_u (=0), discussed in Section 2.9.3. Effective stress analysis is carried out in terms of drained shear strength parameters c' and ϕ'.

2.9.3 Undrained Shear Strength of Clays

The failure envelope of a saturated clay during undrained loading, in terms of *total stresses* is horizontal. Therefore, with subscript u denoting undrained loading, $\phi_u = 0$ and the shear strength $\tau_f = c_u$ at any stress level. Here, c_u is commonly known as the *undrained shear strength* of the clay. It is a total stress parameter that varies with the water content (i.e., the consolidation pressure) of the clay. Larger the consolidation pressure, lesser is the water content and larger is the undrained shear strength.

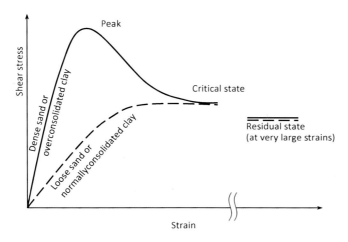

Fig. 2.17 Stress strain plots showing peak, critical and residual states

Short-term analysis of a saturated clay is generally carried out in terms of total stresses using total stress parameters c_u and $\phi_u = 0$. There are a few ways of deriving the undrained shear strength c_u of a clay. They are:

- unconsolidated undrained triaxial test (ASTM D2850; BS1377-7; AS 1289.6.4.1),
- unconfined compression test (ASTM D2166; BS1377-9),
- field (ASTM D2573; BS1377-9; AS 1289.6.2.1) or laboratory (ASTM D4648; BS1377-7) vane,
- handheld torvane,
- laboratory vane (ASTM D4648),
- pocket penetrometer, and
- empirical correlations (Chap. 3).

The ratio of intact to remoulded undrained shear strength, at the same water content, is known as the sensitivity of the clay. It is denoted by S_t and is greater than unity. It is approximately equal to the ratio of the peak to residual undrained shear strength, which can be determined by a vane shear test.

2.9.4 Peak, Residual and Critical States

Figure 2.17 shows typical stress-strain plots of two different types of soils: (a) normally consolidated clays or loose sands, and (b) overconsolidated clays or dense sands. The first one exhibits contractive (i.e. volume reduction) behavior throughout the shearing process, and the shear stress peaks at critical state which occurs at relatively large strains. At critical state, the soil deforms under constant volume or void ratio. The second type of soil exhibits dilative (i.e. volume increase) behavior during the shear after some initial contraction, and the peak value of the shear stress is reached at relatively small strain in the order of 1–2 %, where the friction angle is known as the *peak friction angle* ϕ'_{peak}. The shear stress reduces on further straining, and reaches the critical state which occurs at strains in the order of 10–20 %.

Irrespective of the initial state (i.e. normally consolidated or overconsolidated clay; or lose or dense sand), the shear stress and the void ratio reach a constant value at critical state. The friction angle and the void ratio at the critical state are known as the *critical state friction angle* ϕ'_{cv} and *critical state void ratio* e_{cv}, respectively, where "cv" stands for constant volume or critical state void ratio. ϕ'_{cv} largely depends on the mineralogy of the soil grains.

At very large strains, clay soils would go from critical state to *residual state*, where the friction angle is known as the *residual friction angle* ϕ'_{res}. In clays, strains in the order of 100 % or more are required to remold the clay and reach residual state (Kulhawy and Mayne 1990). In granular soils, residual state is not far from the critical state, where it can be assumed that $\phi'_{res} \approx \phi'_{cv}$. For cohesive soils ϕ'_{res} can be several degrees lower than ϕ'_{cv}. For all soils in general, $\phi'_{res} < \phi'_{cv} < \phi'_{peak}$. ϕ'_{res} and the residual shear strength depend largely on the mineralogy of the grains, grain sizes and shapes, clay fraction, and plasticity.

2.9.5 Dilatancy Angle

In granular soils, *dilatancy* increases with increasing relative density and decreasing confining stress levels. At a specific relative density, the granular soil can exhibit contracting or dilating behavior, depending on the confining pressure. The difference between ϕ'_{peak} and ϕ'_{cv} is the friction angle component that is caused by dilatancy, which is quantified by the *dilatancy angle* ψ defined as

$$\tan \psi = -\frac{d\varepsilon_{vol}}{d\gamma} \qquad (2.60)$$

where, ε_{vol} is the volumetric strain and γ is the shear strain. The dilatancy angle varies during the shear and the maximum which occurs at the peak shear stress is generally used in computations. Bolton (1986) noted that $\phi'_{cv} = 33°$ for quartz sands and 40° for feldspar sands. However, natural sand deposits often contain significant silt content and hence will have a lower ϕ'_{cv}. He suggested that ϕ'_{cv} for most natural sands will rarely be above the range of 30°–33°, and can be as low as 27° when there is high silt content. ϕ'_{cv} values in the range of 27°–37° have been reported in the literature (Bolton 1986; Salgado et al. 2000).

2.9.6 Coefficient of Earth Pressure at Rest

The ratio of effective horizontal normal stress (σ'_h) to the effective vertical normal stress (σ'_v) is known as the coefficient of earth pressure K. When the soil is loaded such that there is no horizontal strain, K is known as the coefficient of earth pressure at rest, denoted by K_0. K_0 can be measured in triaxial compression tests, or in situ using devices such as self-boring pressuremeter, K_0 stepped blade, earth pressure cells, hydraulic fracture, etc.

2.10 Soil Variability

Harr (1977) reported that the coefficient of variation of the compressive strength of concrete and tensile strength of steel are about 6 %. This is not the case with soil parameters. There is significant variability associated with soils and rocks when compared with other engineering materials such as steel or concrete. This variability increases the risk associated with our predictions. Table 2.11 summarises the

Table 2.11 Typical values of coefficient of variation

Parameter	Coeff. of variation (%)
Void ratio, e	20–30
Porosity, n	20–30
Relative density, D_r	10–40
Specific gravity, G_s	2–3
Unit weight, γ	3–7
Buoyant unit weight, γ'	0–10
Liquid limit, LL	10–20
Plastic limit, PL	10–20
Plasticity index, PI	30–70[a]
Optimum water content, w_{opt}	20–40[b]
Maximum dry density, $\rho_{d(max)}$	5
California bearing ratio, CBR	25
Permeability (saturated), k	70–90
Permeability (Unsaturated), k	130–240
Coefficient of consolidation, c_v	25–70
Compression index, C_c	10–40
Preconsolidation pressure, σ'_c	10–35
Friction angle (sands), ϕ'	10
Friction angle (clays), ϕ'	10–50
Friction angle (mine tailings), ϕ'	5–20
Undrained shear strength, c_u	20–40
Standard penetration test blow count, N	20–40
Cone (Electric) resistance, q_c	5–15
Cone (Mechanical) resistance, q_c	15–35
Undrained shear strength from field vane, c_u	10–30

[a]Lower values for clays and higher ones for sandy/gravelly clays
[b]Lower values for clays and higher ones for granular soils

coefficient of variation values of some geotechnical parameters reported in the literature (Duncan 2000; Sivakugan 2011). Here, the coefficient of variation is defined as

$$\text{Coefficient of variation } (\%) = \frac{\text{Standard deviation}}{\text{Mean}} \times 100 \qquad (2.61)$$

References

Alonso EE, Gens A, Lloret A (2000) Precompression design for secondary settlement reduction. Geotechnique 50(6):645–656
AS 1289.2.1.1-2005 Methods of testing soils for engineering purposes, Method 2.1.1: Soil moisture content tests – determination of the moisture content of a soil – oven drying method (standard method), Australian Standard
AS 1289.3.1.1-2009 Methods of testing soils for engineering purposes, Method 3.1.1: Soil classification tests – determination of the liquid limit of a soil – four point Casagrande method, Australian Standard
AS 1289.3.2.1-2009 Methods of testing soils for engineering purposes, Method 3.2.1: Soil classification tests – determination of the plastic limit of a soil – standard method, Australian Standard
AS 1289.3.5.1-2006 Methods of testing soils for engineering purposes, Method 3.5.1: Soil classification tests – determination of the soil particle density of a soil – standard method, Australian Standard
AS 1289.3.6.1-2009 Methods of testing soils for engineering purposes, Method 3.6.1: Soil classification tests – determination of the particle size distribution of a soil – standard method of analysis by sieving, Australian Standard
AS 1289.3.6.3-2003 Method of testing soils for engineering purposes, Method 3.6.3: Soil classification tests – determination of the particle size distribution of a soil – standard method of fine analysis using a hydrometer, Australian Standard
AS 1289.3.9.1-2002 Methods of testing soils for engineering purposes, Method 3.9.1: Soil classification tests – determination of the cone liquid limit of a soil, Australian Standard
AS 1289.5.1.1-2003. Methods of testing soils for engineering purposes, Method 5.1.1: Soil compaction and density tests – determination of the dry density/moisture content relation of a soil using standard compactive effort, Australian Standard
AS 1289.5.2.1-2003 Methods of testing soils for engineering purposes, Method 5.2.1: Soil compaction and density tests – determination of the dry density/moisture content relation of a soil using modified compactive effort, Australian Standard
AS 1289.5.5.1-1998 Methods of testing soils for engineering purposes, Method 5.5.1: Soil compaction and density tests – determination of the minimum and maximum dry density of a cohesionless material – standard method, Australian Standard
AS 1289.6.2.1-2001 Methods of testing soils for engineering purposes, Method 6.2.1: Soil strength and consolidation tests – determination of the shear strength of a soil – field test using a vane, Australian Standard
AS 1289.6.4.1-1998. Methods of testing soils for engineering purposes, Method 6.4.2: Soil strength and consolidation tests – determination of the compressive strength of a soil – compressive strength of a saturated specimen tested in undrained triaxial compression without measurement of pore water pressure, Australian Standard

References

AS 1289.6.6.1-1998 Methods of testing soils for engineering purposes, Method 6.6.1: Soil strength and consolidation tests – determination of the one-dimensional consolidation properties of soil – standard method, Australian Standard

AS 1289.6.7.1-2001 Methods of testing soils for engineering purposes, Method 6.7.1: Soil strength and consolidation tests – determination of permeability of a soil – constant head method for a remoulded specimen, Australian Standard

AS 1289.6.7.2-2001 Methods of testing soils for engineering purposes, Method 6.7.2: Soil strength and consolidation tests – determination of permeability of a soil – falling head method for a remoulded specimen, Australian Standard

AS 4678-2002 Earth retaining structures, Australian Standard

ASTM D2435-04 Standard methods for one-dimensional consolidation properties of soils using incremental loading

ASTM D1557-09 Standard test method for laboratory compaction characteristics of soil using modified effort (56,000 ft-lbf/ft^3 (2,700 kN-m/m^3))

ASTM D2166-06 Standard test method for unconfined compressive strength of cohesive soil

ASTM D2216-05. Standard test methods for laboratory determination of water (moisture) content of soil and rock by mass.

ASTM D2434-68 Standard test method for permeability of granular soils (Constant head)

ASTM D2573-08 Standard test method for field vane shear test in cohesive soil

ASTM D2850-03a Standard test method for unconsolidated-undrained triaxial compression test on cohesive soils

ASTM D422-63 Standard test method for particle-size analysis of soils

ASTM D4253-00 Standard test method for maximum index density and unit weight of soils using a vibrating table

ASTM D4254-00 Standard test method for minimum index density and unit weight of soils and calculation of relative density

ASTM D4318-10 Standard test methods for liquid limit, plastic limit and plasticity index of soils

ASTM D4648-10 Standard test method for laboratory miniature vane shear test for saturated fine grained clayey soil

ASTM D5084-03 Standard test methods for measurement of hydraulic conductivity of saturated porous materials using a flexible wall permeameter

ASTM D6913-04 Standard test methods for particle-size distribution (gradation) of soils using sieve analysis

ASTM D698-07 Standard test method for laboratory compaction characteristics of soil using standard effort (12400 ft-lbf/ft^3 (600 kN-m/m^3))

ASTM D854-06e1 Standard test methods for specific gravity of soil solids by water pycnometer

Balasubramaniam AS, Cai H, Zhu D, Surarak C, Oh EYN (2010) Settlements of embankments in soft soils. Geotech Eng J SEAGS & AGSSEA 41(2), 19 p

Bell FG (2000) Engineering properties of soils and rocks, 4th edn. Blackwell Science, London

Bolton MD (1986) The strength and dilatancy of sands. Geotechnique 36(1):65–78

BS 1377 (1990) Methods of test for soils for civil engineering purposes. British Standard Institution, London

Burmister DM (1949) Principles and techniques of soil identification. In: Proceedings of annual highway research board meeting, National Research Council, Washington, DC, 29, pp 402–433

Casagrande A (1936) The determination of the pre-consolidation load and its practical significance, Discussion D-34. In: Proceedings of the first international conference on soil mechanics and foundation engineering, Cambridge, III, pp 60–64

Casagrande A (1948) Classification and identification of soils. ASCE Trans 113:901–930

Casagrande A, Poulos SJ (1969) On the effectiveness of sand drains. Can Geotech J 6(3):287–326

Cedegren HR (1967) Seepage, drainage and flow nets. Wiley, New York

D'Arcy H (1856) Les fontaines publiques de la ville de Dijon (The public fountains of the city of Dijon). Dalmont, Paris

Das BM (2009) Soil mechanics laboratory manual, 7th edn. Oxford University Press, Oxford

Das BM (2010) Principles of geotechnical engineering, 7th edn. Cengage Learning, Boston, 666 pp

Domenico PA, Mifflin MD (1965) Water from low-permeability sediments and land subsidence. Water Resour Res Am Geophys Union 1(4):563–576

Duncan JM (2000) Factor of safety and reliability in geotechnical engineering. J Geotech Geoenviron Eng ASCE 126(4):307–316

Fukushima S, Ishii T (1986) An experimental study of the influence of confining pressure on permeability coefficients of fill dam core materials. Soils Found 26(4):32–46

Handy RL, Spangler MG (2007) Geotechnical engineering: soil and foundation principles and practice, 5th edn. McGraw Hill, New York, 904 pp

Harr ME (1962) Ground water and seepage. McGraw Hill, New York

Harr ME (1977) Mechanics of particulate media. McGraw Hill, New York

Hausmann M (1990) Engineering principles of ground modification. McGraw Hill, Singapore, 632 p

Holtz RD, Kovacs WD, Sheehan TC (2011) An introduction to geotechnical engineering, 2nd edn. Pearson, London, 863 pp

Kenney TC, Lau D, Ofoegbu GI (1984) Permeability of compacted granular materials. Can Geotech J 21(4):726–729

Kulhawy FH, Mayne PW (1990) Manual on estimating soil properties for foundation design, Report EL-6800. Electric Power Research Institute, Palo Alto

Lambe TW, Whitman RV (1979) Soil mechanics SI version. Wiley, New York, 553 p

McCarthy DF (1977) Essentials of soil mechanics and foundations. Prentice Hall, Reston, 505 p

Mitchell JK (1976) Fundamentals of soil behavior. Wiley, New York

Muskat M (1946) The flow of homogeneous fluids through porous media. J.W. Edwards, Ann Arbor

Salgado R, Bandini P, Karim A (2000) Shear strength and stiffness of silty sand. J Geotech Geoenviron Eng ASCE 126(5):451–462

Scheidegger AE (1957) The physics of flow through porous media. Macmillan, New York

Schmertmann JH (1955) The undisturbed consolidation behavior of clays. Trans ASCE 120:1201–1233

Skempton AW (1953) The colloidal activities of clays. In: Proceedings of the 3rd ICSMFE, I, Zurich, pp 57–61

Sivakugan N (2011) Chapter 1: Engineering properties of soil. In: Das BM (ed) Geotechnical engineering handbook. J. Ross Publishing, Ft. Lauderdale

Sivakugan N, Das BM (2010) Geotechnical engineering: a practical problem solving approach. J. Ross Publishing, Ft. Lauderdale, 506 pp

Sivakugan N, Vigneswaran B (1991) A simple analysis of constant rate loading in clays. In: Proceedings of the 13th Canadian congress of applied mechanics, Manitoba, pp 790–791

Sivakugan N, Arulrajah A, Bo MW (2011) Laboratory testing of soils, rocks and aggregates. J. Ross Publishing, Ft. Lauderdale

Sowers GB, Sowers GF (1961) Introductory soil mechanics and foundations, 2nd edn. The Macmillan Co., New York, 386 p

Tavenas F, Leroueil S (1987) State-of-the-art on laboratory and in-situ stress-strain-time behavior of soft clays. In: Proceedings of international symposium on geotechnical engineering of soft soils, Mexico City, pp 1–46

Terzaghi K (1925) Erdbaumechanik auf Bodenphysikalischer Grundlage. Franz Deuticke, Leipzig und Wein, 399

Terzaghi K (1943) Theoretical soil mechanics. Wiley, New York

Terzaghi K, Peck R (1967) Soil mechanics in engineering practice, 2nd edn. Wiley, New York

U.S. Navy (1982) Soil mechanics – design manual 7.1, Department of the Navy, Naval Facilities Engineering Command, U.S. Government Printing Office, Washington, DC

Winchell AN (1942) Elements of mineralogy. Prentice-Hall, Englewood Cliffs

Wu TH, Chang NY, Ali EM (1978) Consolidation and strength properties of a clay. J Geotech Eng Div ASCE 104(GT7):899–905

Wong PK (2006) Preload design, part 1: review of soil compressibility behavior in relation to the design of preloads. Australian Geomechanics Society Sydney chapter, 2006 symposium, pp 27–32

Chapter 3
Correlations for Laboratory Test Parameters

Abstract With the necessary theoretical framework covered in Chapter 2, this chapter discusses the correlations relating the different soil parameters determined in the laboratory for both cohesive and cohesionless soils. Parameters covered in this chapter include permeability, consolidation, undrained and drained shear strength, stiffness and modulus and coefficient of earth pressure at rest. The relationships between the parameters discussed herein are not necessarily all empirical. Some theoretical relationships are also given. In addition to the theoretical and empirical relationships, typical values of the parameters are provided wherever possible. Correlations with laboratory data to be directly used in pile design are also provided.

Keywords Laboratory tests • Design parameters • Correlations • Consolidation • Shear strength • Pile design

With the necessary theoretical framework covered in Chap. 2, this chapter discusses the correlations relating the different soil parameters determined in the laboratory. The relationships between the parameters discussed herein are not necessarily all empirical. Some theoretical relationships are also given. In addition to the theoretical and empirical relationships, typical values of the parameters are provided wherever possible.

3.1 Permeability

Permeability relationships for granular and cohesive soils depend on different parameters. Therefore, they are covered separately in this section.

3.1.1 Granular Soils

Granular soils have higher permeability than cohesive soils. Within granular soils, the permeability increases with the grain size. Generally, granular soils are assumed to be free draining. However, when they contain more than 15 % fines, they are no longer free draining. Fines in excess of 30 % can reduce the permeability significantly.

In clean uniform loose sands with less than 5 % fines, with D_{10} in the range of 0.1–3.0 mm, Hazen (1911, 1930) suggested that the permeability k can be related to D_{10} by

$$k(\text{cm/s}) = C\, D_{10}^2 \tag{3.1}$$

where D_{10} is in mm, and C is a constant that varies between 0.5 and 1.5. The scatter in C is considerably large as reported by many researchers and documented by Carrier III (2003), who suggested using Kozeny-Carman equation instead of Hazen's.

Kozeny-Carman equation, proposed by Kozeny (1927) and improved by Carman (1938, 1956) is:

$$k = \frac{1}{C_{K-C} S^2} \frac{\gamma_w}{\mu_w} \frac{e^3}{1+e} \tag{3.2}$$

where C_{K-C} = Kozeny-Carman factor (approximately 5) to account for the pore shape and tortuosity of the flow channels, and S = specific surface area per unit *volume* of grains. For uniform spherical grains, $S = 6/D$ where D is the grain diameter. For non-spherical grains of different sizes, determining S is not straightforward. Carrier III (2003) modified Eq. (3.2) slightly and suggested a method to derive the specific surface from the sieve analysis data.

Lambe and Whitman (1979) noted that e versus $\log k$ variation is often linear for both fine and coarse grained soils. Further, k varies linearly with e^2, $e^2/(1+e)$, and $e^3/(1+e)$ in granular soils. Figure 3.1 shows the k-e-D_{10} chart proposed by US Navy (1982) for coarse grained soils with $C_u = 2$–12 and $D_{10}/D_5 < 1.4$, based on laboratory test data on remolded compacted sand. Chapuis (2004) related k (cm/s), e and D_{10} (mm) through the following equation for natural uniform sand and gravel, which is valid when permeability is in the range of 10^{-1}–10^{-3} cm/s.

$$k(\text{cm/s}) = 2.4622 \left[D_{10}^2 \left(\frac{e^3}{1+e} \right) \right]^{0.7825} \tag{3.3}$$

k-e-D_{10} variation, based on Chapuis (2004) equation are also shown in Fig. 3.1 for comparison.

3.1 Permeability

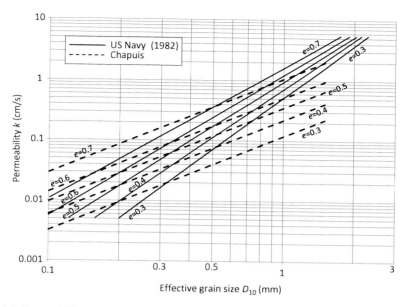

Fig. 3.1 Permeability – void ratio – effective grain size relation for coarse grained soils from US Navy (1982) and Chapuis (2004)

3.1.2 Cohesive Soils

Kozeny-Carman equation does not hold very well for cohesive soils. However, as noted by Taylor (1948) and Lambe and Whitman (1979), void ratio is proportional to the logarithm of permeability. Therefore,

$$\log k = \log k_0 - \frac{e_0 - e}{C_k} \tag{3.4}$$

where, k is the permeability at void ratio of e (possibly under some surcharge), and k_0 is the in situ permeability at in situ void ratio of e_0. $1/C_k$ is the slope of the log k versus e line. C_k is the dimensionless *permeability change index* that can be taken as approximately 0.5 e_0. Equation (3.4) works well for $e_0 < 2.5$. Mesri and Olsen (1971) suggested that, for clays, log k varies linearly with log e.

For remoulded clays, Carrier III and Beckman (1984) showed that

$$k(\text{m/s}) \approx \frac{0.0174}{1+e} \left\{ \frac{e - 0.027(PL - 0.242PI)}{PI} \right\}^{4.29} \tag{3.5}$$

The permeability of compacted clays is significantly lower for clays compacted wet of optimum than dry of optimum. For applications requiring low permeability (e.g., clay liners at the bottom of waste disposal ponds), it may be better to compact at water contents greater than the optimum water content.

3.2 Consolidation

There are several parameters defining the consolidation behavior of clays. They include, compression index, recompression index, constrained modulus, coefficient of consolidation and coefficient of secondary compression. They are discussed separately in this section.

3.2.1 Compression Index

Compression index C_c (see Fig. 2.12b) is the slope of the *virgin consolidation line*, a straight line in the $e - \log\sigma'_v$ space. The $e - \sigma'_v$ values will be located on this line when the clay is normally consolidated, irrespective of the stress level. C_c is a measure of how stiff the clay is when it is normally consolidated, and is an important parameter in computing the final consolidation settlements. It is often related to the in situ natural water content w_n, initial in situ void ratio e_0, liquid limit *LL*, or plasticity index *PI*. Kulhawy and Mayne (1990) suggested that the correlations based on natural water content work better than the ones based on *LL* or e_0. Koppula (1981) evaluated the relationship between C_c and eight other parameters and observed that the one with the least error is given by

$$C_c = 0.01 w_n \tag{3.6}$$

where, w_n is in percentage. For saturated soils, assuming $G_s = 2.70$, Eq. (3.6) can be written as

$$C_c = 0.37 e_0 \tag{3.7}$$

Kulhawy and Mayne (1990) classify the clays based on compressibility as shown in Table 3.1.

Winterkorn and Fang (1975) tabulated C_c values showing that they are significantly larger for undisturbed clays than the remolded ones. Some of the empirical correlations for C_c are summarized in Table 3.2. Further correlations for C_c are given in Sridharan and Nagaraj (2000) and Djoenaidi (1985). Some typical values of compression index reported in the literature are summarized in Table 3.3.

3.2 Consolidation

Table 3.1 Compressibility classification based on C_c

Compressibility	C_c
Slight or low	<0.2
Moderate of intermediate	0.2–0.4
High	>0.4

Table 3.2 Empirical correlations for C_c

Correlation	Comments	References
$C_c = 0.009\ (LL\text{-}10)$	Undisturbed clay of sensitivity less than 4. Reliability ± 30 %	Terzaghi and Peck (1948)
$C_c = 0.007\ (LL\text{-}10)$	Remoulded clay	Skempton (1944)
$C_c = 0.0046\ (LL\text{-}9)$	Sao Paulo, Brazil clays	Cozzolino (1961)
$C_c = 0.0186\ (LL\text{-}30)$	Soft silty Brazilian clays	Cozzolino (1961)
$C_c = 0.01\ (LL\text{-}13)$	All clays	USACE (1990)
$C_c = 0.008\ (LL\text{-}8.2)$	Indiana soils	Lo and Lovell (1982)
$C_c = 0.21 + 0.008\ LL$	Weathered & soft Bangkok clays	Balasubramaniam and Brenner (1981)
$C_c = 0.30\ (e_0 - 0.27)$	Inorganic silty clay	Hough (1957)
$C_c = 1.15(e_0 - 0.35)$	All clays	Azzouz et al. (1976)
$C_c = 0.75(e_0 - 0.50)$	Soils of very low plasticity	Azzouz et al. (1976)
$C_c = 0.4(e_0 - 0.25)$	Clays from Greece & parts of US	Azzouz et al. (1976)
$C_c = 0.141 G_s^{1.2} \left(\frac{1+e_0}{G_s}\right)^{2.382}$	90 samples; Bowles (1988) suggests e_0 be less than 0.8	Rendon-Herrero (1980)
$C_c = 0.256 + 0.43(e_0 - 0.84)$	Brazilian clays	Cozzolino (1961)
$C_c = 0.54\ (e_0 - 0.35)$	All clays	Nishida (1956)
$C_c = 0.22 + 0.29\ e_0$	Weathered and soft Bangkok clays	Balasubramaniam and Brenner (1981)
$C_c = 0.575\ e_0 - 0.241$	French clays	Balasubramaniam and Brenner (1981)
$C_c = 0.5363(e_0 - 0.411)$	Indiana soils	Goldberg et al. (1979)
$C_c = 0.496\ e_0 - 0.195$	Indiana soils	Lo and Lovell (1982)
$C_c = 0.40(e_0 - 0.25)$	Clays from Greece & parts of US	Azzouz et al. (1976)
$C_c = 0.01\ w_n$	Chicago clays	Azzouz et al. (1976)
$C_c = 0.01\ w_n$	Canada clays	Koppula (1981)
$C_c = 0.0115\ w_n$	Organic soils, peat	USACE (1990) and Azzouz et al. (1976)
$C_c = 0.012\ w_n$	All clays	USACE (1990)
$C_c = 0.01(w_n - 5)$	Clays from Greece & parts of US	Azzouz et al. (1976)
$C_c = 0.0126\ w_n - 0.162$	Indiana soils	Lo and Lovell (1982)
$C_c = 0.008\ w_n + 0.20$	Weathered soft Bangkok clays	Balasubramaniam and Brenner (1981)
$C_c = 0.0147\ w_n - 0.213$	French clays	Balasubramaniam and Brenner (1981)
$C_c = (1 + e_0)[0.1 + 0.006(w_n - 25)]$	Varved clays	USACE (1990)

Table 3.3 Typical values of compression index for undisturbed clays

Soil	C_c	References
Normally consolidated medium sensitive clays	0.2–0.5	Holtz and Kovacs (1981)
Organic silt and clayey silts (ML-MH)	1.5–4.0	
Organic clays (OH)	>4	
Peats (Pt)	10–15	
Boston blue clay, undisturbed (CL)	0.35	Lambe and Whitman (1979)
Chicago clay undisturbed (CH)	0.42	
Cincinnati clay (CL)	0.17	
Louisiana clay, undisturbed (CH)	0.33	
New Orleans clay undisturbed (CH)	0.29	
Siburua clay (CH)	0.21	
Kaolinite (CL/CH)	0.21–0.26	
Na-Montmorillonite(CH)	2.6	
Chicago silty clay (CL)	0.15–0.30	Holtz and Kovacs (1981)
Boston blue clay (CL)	0.3–0.5	
Vicksburg buckshot clay (CH)	0.5–0.6	
Swedish medium sensitive clays (CL-CH)	1–3	
Canadian Leda clays (CL-CH)	1–4	
Mexico City clay (MH)	7–10	
San Francisco Bay mud (CL)	0.4–1.2	
Bangkok clays (CH)	0.4	
Uniform sand, loose (SP)	0.05–0.06	USACE (1990)
Uniform sand, dense (SP)	0.02–0.03	
Uniform silts (ML)	0.2	

For normally consolidated clays, m_v and C_c are related by

$$m_v = \frac{0.434 C_c}{(1+e_0)\sigma'_{average}} \quad (3.8)$$

where $\sigma'_{average}$ is the average value of the vertical normal stress during consolidation.

The undrained shear strengths of a clay at plastic limit and liquid limit are approximately 170 kPa and 1.7 kPa, respectively, differing by about 100 times. Noting that the undrained shear strength is proportional to the effective consolidation pressures, the effective consolidation pressures at plastic limit and liquid limit also would differ by 100 times. Noting that the change in void ratio between the plastic and liquid limit of a saturated clay is given by $PI \times G_s$, the compression index can be written as (Wroth and Wood 1978)

$$C_c = G_s \left(\frac{PI}{200} \right) \quad (3.9)$$

3.2.2 Recompression Index or Swelling Index

There are no reliable correlations reported in the literature for the recompression index (C_r) or the swelling index (C_s), which can be assumed to be equal for all practical purposes. In reality, the recompression index can be slightly less than the swelling index.

Recompression index can be estimated on the basis that C_r/C_c is typically in the range of 1/5–1/10. There are exceptions. Lambe and Whitman (1979) reported that in Na-Montmorillonite, the swelling index can be as high as 2.5.

During recompression, m_v and C_r are related by

$$m_v = \frac{0.434 C_r}{(1+e_0)\sigma'_{average}} \tag{3.10}$$

where $\sigma'_{average}$ is the average value of the vertical normal stress during consolidation while the clay is still overconsolidated.

In critical state soil mechanics, the stress path is monitored in the three dimensional $\ln p'$-q-V space. Here p' is the mean effective stress, defined as $(\sigma'_1 + \sigma'_2 + \sigma'_3)/3$, q is the deviator stress defined as $\sigma_1 - \sigma_3$, and $V =$ specific volume defined as $1+e$. The parameters λ and κ, very similar to C_c and C_r, are the slopes of the virgin consolidation line and the unloading line in the $V - \ln p'$ space where the specific volume $V (=1+e)$ is plotted against the natural logarithm of the mean effective stress p'. It can be shown that

$$C_c = \lambda \ln 10 = 2.3026\lambda \tag{3.11}$$
$$C_r = \kappa \ln 10 = 2.3026\kappa \tag{3.12}$$

The *plastic volumetric strain ratio* Λ is defined as

$$\Lambda = \frac{\lambda - \kappa}{\lambda} \tag{3.13}$$

It is a parameter that is commonly used in the critical state soil mechanics.

3.2.3 Compression Ratio and Recompression Ratio

In the early days of soil mechanics, a parameter known as *compression ratio* (*CR*) or *modified compression index* ($C_{c\varepsilon}$) was used widely in computing consolidation settlements. It is similar to C_c, and is the slope of the virgin compression line when the *vertical normal strain* (instead of void ratio) is plotted against the logarithm of effective normal stress. It is defined as $C_c/(1+e_0)$ where e_0 is the initial void ratio. For most clays subjected to consolidation tests, it varies in the range of 0.2–0.4.

Table 3.4 Classification based on soil compressibility CR or RR

Description	$\frac{C_c}{1+e_0}$ or $\frac{C_r}{1+e_0}$
Very slightly compressible	<0.05
Slightly compressible	0.05–0.10
Moderately compressible	0.10–0.20
Highly compressible	0.20–0.35
Very highly compressible	>0.35

Table 3.5 Empirical correlations for the compression ratio

Soil type	Correlation	References
Marine clays of southeast Asia	$CR = 0.0043\ w_n$	Azzouz et al. (1976)
	$CR = 0.0045\ LL$	Balasubramaniam and Brenner (1981)
Bangkok clays	$CR = 0.00463 LL - 0.013$	Balasubramaniam and Brenner (1981)
	$CR = 0.00566\ w_n - 0.037$	
French clays	$CR = 0.0039\ w_n + 0.013$	Balasubramaniam and Brenner (1981)
Indiana clays	$CR = 0.0249 + 0.003\ w_n$	Lo and Lovell (1982)
	$CR = 0.0294 + 0.00238\ LL$	
	$CR = 0.0125 + 0.0.152 e_0$	
Indiana clays	$CR = 0.2037(e_0 - 0.2465)$	Goldberg et al. (1979)
Clays from Greece & parts of US	$CR = 0.002\ (LL + 9)$	Azzouz et al. (1976)
	$CR = 0.14(e_0 + 0.007)$	
	$CR = 0.003\ (w_n + 7)$	
	$CR = 0.126(e_0 + 0.003 LL\text{-}0.06)$	
Chicago clays	$CR = 0.208\ e_0 + 0.0083$	Azzouz et al. (1976)
Inorganic & organic clays and silty soils	$CR = 0.156\ e_0 + 0.0107$	Elnaggar and Krizek (1970)

Similarly, a *recompression ratio* (*RR*) or a *modified recompression index* is defined as $C_r/(1 + e_0)$. Based on the compression ratio or recompression ratio, the compressibility of a clay can be classified as shown in Table 3.4. *CR* and *RR* are still used by the practicing engineers.

Selected empirical correlations for the compression ratio, from the extensive list, collated and presented by Djoenaidi (1985) are listed in Table 3.5.

3.2.4 Constrained Modulus

The constrained modulus D is approximately related to the preconsolidation pressure by (Canadian Geotechnical Society 1992):

$$D = (40 \text{ to } 80)\sigma'_p \qquad (3.14)$$

where, the upper end of the range is applicable for stiff clays and lower end for the soft clays.

3.2.5 Coefficient of Consolidation c_v

When a load is applied at the ground level, how quickly the consolidation process is completed depends on the coefficient of consolidation c_v. Larger the c_v, faster is the consolidation process. Generally, c_v is an order of magnitude larger in overconsolidated clays than in normally consolidated clays. It can be deduced from Eq. (2.39) that c_v increases with increasing permeability and stiffness of the soil skeleton. Stiffer soil skeletons enables faster consolidation.

c_v can vary from less than 1 m²/year for low permeability clays to as high as 1000 m²/year for sandy clays of high permeability. Tezaghi et al. (1996) suggested that clays with $LL = 10$–100 have c_v in the range of 0.3–30 m²/year. Figure 3.2 proposed by U.S. Navy (1982) can be used as a rough guide or first order estimates for checking c_v values determined in the laboratory. Soil disturbance delays the consolidation and hence reduces the coefficient of consolidation of both normally consolidated and overconsolidated clays.

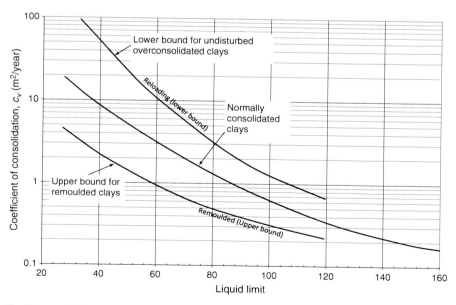

Fig. 3.2 $c_v - LL$ relation (After U.S. Navy 1982)

Table 3.6 Some typical C_α/C_c values

Material	C_α/C_c
Granular soils including rockfill	0.02 ± 0.01
Shale and mudstone	0.03 ± 0.01
Inorganic clays and silts	0.04 ± 0.01
Organic clays and silts	0.05 ± 0.01
Peats and muskeg	0.06 ± 0.01

After Mesri et al. (1994)

3.2.6 Secondary Compression

Mesri and Godlewski (1977) suggested that the ratio of C_α/C_c generally varies in the range of 0.025–0.10, with an average value of about 0.05. The upper end of the range applies to organic soils including peat and muskeg. The lower end is for inorganic soils including clays and granular soils. Some of the values suggested by Mesri et al. (1994) are given in Table 3.6.

As a first approximation, $C_{\alpha\varepsilon}$ [see Eq. (2.56)] of normally consolidated clays can be estimated as (US Navy 1982)

$$C_{\alpha\varepsilon} = 0.0001 w_n \text{ for } 10 < w_n(\%) < 3000 \qquad (3.15)$$

where w_n is the natural water content in percentage. When overconsolidated, C_α and $C_{\alpha\varepsilon}$ can be significantly less, in the order of 30–50 % of the values reported for normally consolidated clays. Figure 3.3 can be used for estimating the modified secondary compression index from the natural water content, for normally consolidated clays.

On the basis of the coefficient of secondary compression clays can be classified as shown in Table 3.7.

3.3 Shear Strength Parameters c' and ϕ'

The cohesion c and friction angle ϕ are the two main shear strength parameters required in any geotechnical analysis. They are discussed in this section, along with relevant empirical correlations. The different ways of defining these parameters and their inter-relationships are discussed here.

3.3.1 Cohesion in Terms of Effective Stress c'

In terms of effective stresses, the failure envelope generally passes through the origin in the τ-σ' plane for most normally consolidated soils, suggesting $c' = 0$. Only in the case of cemented soils, partially saturated soils and heavily overconsolidated

3.3 Shear Strength Parameters c' and ϕ'

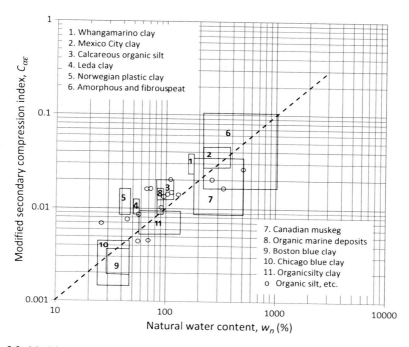

Fig. 3.3 Modified secondary compression index versus natural water content for NC clays (Adapted from Holtz and Kovacs 1981; Data from Mesri 1973)

Table 3.7 Classification based on $C_{\alpha\varepsilon}$

Description	$C_{\alpha\varepsilon}$
Very low	<0.002
Low	0.002–0.004
Medium	0.004–0.008
High	0.008–0.016
Very high	0.016–0.032
Extremely high	0.064

soils, there can be some effective cohesion. For uncemented soils including clays, the shear strength in terms effective stresses is mainly frictional. Based on the Danish code of practice for foundations, Sorensen and Okkels (2013) suggest that a cautious estimate of c' for overconsolidated clays can be obtained from

$$c' = 0.1 \, c_u \qquad (3.16)$$

They also suggest that c' is poorly correlated to *PI*. Australian Standards for retaining walls (AS 4678) suggests the values for c' and ϕ' in Table 3.8.

Table 3.8 Typical values of c' and ϕ'

Soil group	Typical soils in group	Soil parameters c' (kPa)	ϕ' (degrees)
Poor	Soft and firm clay of medium to high plasticity; silty clays; loose variable clayey fills; loose sandy silts	0–5	17–25
Average	Stiff sandy clays; gravelly clays; compact clayey sands and sandy silts; compacted clay fills	0–10	26–32
Good	Gravelly sands, compacted sands, controlled crushed sandstone and graveled fills, dense well graded sands	0–5	32–37
Very good	Weak weathered rock, controlled fills of road base, gravel and recycled concrete	0–25	36–43

After AS 4678-2002

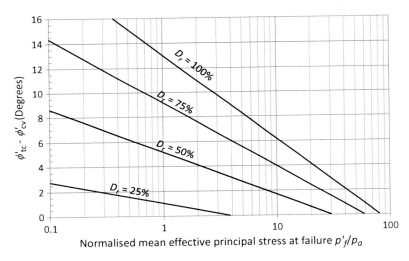

Fig. 3.4 Dilatancy angle from triaxial compression tests versus normalised mean effective stress for different relative densities in sands (Adapted from Bolton 1986)

For natural intact normally consolidated clays, ϕ' can vary from less than 20° to little more than 30°. For compacted clays, ϕ' is typically in the range of 25°–30°, but can be slightly higher.

3.3.2 Effects of Dilatancy in Granular Soils

The relation between the dilatancy component $\phi'_{peak} - \phi'_{cv}$ from a *triaxial compression test*, the relative density, and the mean principal stress at failure p'_f suggested by Bolton (1986) for sands is shown in Fig. 3.4. Here, p_a is the

3.3 Shear Strength Parameters c' and ϕ'

atmospheric pressure, which is about 101.3 kPa. The relationship can be expressed as (Bolton 1986; Kulhawy and Mayne 1990)

$$\phi'_{tc} - \phi'_{cv} = 3\left\{D_r\left[10 - \ln\left(100 \times \frac{p'_f}{p_a}\right)\right] - 1\right\} \quad (3.17)$$

Kulhawy and Mayne (1990) suggest taking $\phi'_{peak} - \phi'_{cv}$ as the dilatancy angle ψ. Bolton (1986) suggested from laboratory test data that for plane strain compression loading

$$\phi'_{peak} = \phi'_{cv} + 0.8\,\psi \quad (3.18)$$

For triaxial compression loading, Eq. 3.18 can be modified as (Salgado 2008)

$$\phi'_{peak} \approx \phi'_{cv} + 0.5\,\psi \quad (3.19)$$

A simple and somewhat crude approximation for dilatancy angle, as often used in Plaxis analysis, is

$$\psi = \phi'_{peak} - 30 \quad (3.20)$$

where $\psi = 0$ for $\phi'_{peak} < 30°$.

Now that we have defined different friction angles, which one should we use in practice? It depends on the level of strain expected in the field situation. Most geotechnical problems involve small strains, and it is unlikely that the peak is exceeded. Therefore, it is recommended to use ϕ'_{peak} as default value. For problems involving large strains ϕ'_{cv} and for those with very large strains (e.g., landslides, slopes, pre-existing shear failures such as old landslide sites) ϕ'_{res} would be appropriate.

3.3.3 ϕ'_{peak}, ϕ'_{cv}, ϕ'_{res} Relationships with Plasticity Index for Clays

There is clear evidence that increasing plasticity leads to a reduction in the peak friction angle ϕ'_{peak}. The increasing plasticity is often due to the increasing clay fraction of flaky grains which have lesser frictional resistance. From the limited data reported in the literature, U.S. Navy (1971) and Ladd et al. (1977) observed the trend between ϕ'_{peak} and plasticity index, shown in Fig. 3.5, for normally consolidated clays, as documented by Holtz and Kovacs (1981). These ϕ'_{peak} values were measured at failure conditions defined as maximum values of σ_1/σ_3 in triaxial

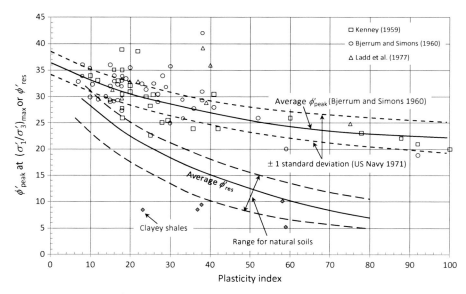

Fig. 3.5 Variation of ϕ'_{peak} and ϕ'_{res} with plasticity index for normally consolidated clays

compression tests. The average values and the ±1 standard deviation band are shown in the figure, along with the test data used in developing these trend lines. It is clear that the peak friction angle decreases with increasing *PI*.

The lower part of Fig. 3.5 shows the variation of the average residual friction angle ϕ'_{res} of normally consolidated cohesive soils with the plasticity index, as suggested by U.S. Air Force (1983). Some test data for clayey shales are also shown in the figure. At residual state, the clays are completely remolded and have undergone very large strains. The clay fraction and the mineralogy are the two factors that govern the residual friction angle ϕ'_{res}. It can range from 15° for kaolinite to 5° for montmorillonite, with illite at 10° (Kulhawy and Mayne 1990). Soils with less than 15 % fines behave like granular soils, with ϕ'_{res} greater than 25°, and close to their ϕ'_{cv}.

Sorensen and Okkels (2013) analysed an extensive database of normally consolidated reconstituted and undisturbed natural clays from the Danish Geotechnical Institute, along with the data from Kenney (1959), Brooker and Ireland (1965), Bjerrum and Simons (1960) and Tezaghi et al. (1996) shown in Fig. 3.6. They suggested that for a cautious lower bound estimate, the peak friction angle can be taken as

$$\phi'_{peak} = 39 - 11 \log PI \qquad (3.21)$$

The best estimate (i.e., mean) of the peak friction angle is given by

3.3 Shear Strength Parameters c' and ϕ'

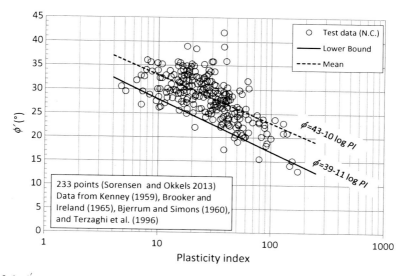

Fig. 3.6 ϕ'_{peak} versus PI for normally consolidated clays (After Sorensen and Okkels 2013)

$$\phi'_{peak} = 43 - 10 \log PI \qquad (3.22)$$

For overconsolidated clays, Sorensen and Okkels (2013) suggested that the cautious lower bound estimate of the peak friction angle can be given by

$$\phi'_{peak} = 44 - 14 \log PI \text{ for } 4 < PI < 50 \qquad (3.23)$$

$$\phi'_{peak} = 30 - 6 \log PI \text{ for } 50 \leq PI < 150 \qquad (3.24)$$

The best estimates for overconsolidated clays are given by

$$\phi'_{peak} = 45 - 14 \log PI \text{ for } 4 < PI < 50 \qquad (3.25)$$

$$\phi'_{peak} = 26 - 3 \log PI \text{ for } 50 \leq PI < 150 \qquad (3.26)$$

Sorensen and Okkels (2013) lower bound estimates for normally consolidated clays are very close to those of overconsolidated clays.

The critical state friction angle ϕ'_{cv} in normally consolidated cohesive soils can be related to PI by (Mitchell 1976; Kulhawy and Mayne 1990)

$$\sin \phi'_{cv} \approx 0.8 - 0.094 \ln PI \qquad (3.27)$$

The data used in developing this relation is shown in Fig. 3.7. The critical state friction angle decreases with increasing PI and activity of the clay mineral. It is greater for Kaolinite of low activity than Montmorillonite of very high activity.

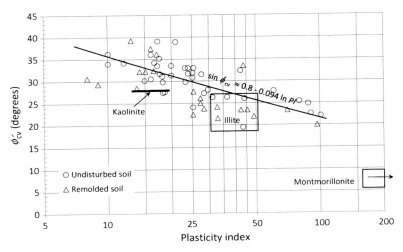

Fig. 3.7 Variation of ϕ'_{cv} with plasticity index for normally consolidated clays

With considerable scatter seen in the Figs. 2.22, 2.23, and 2.24, Eqs. (3.21), (3.22), (3.23), (3.24), (3.25), (3.26), and (3.27) should be used with caution. For normally consolidated clays, ϕ'_{peak} should be very close to the ϕ'_{cv}, which can also be seen from Figs. 3.5 and 3.7.

3.3.4 Other Friction Angle Correlations

What is referred to simply as friction angle in literature, especially in most textbooks, generally means the peak friction angle in terms of effective stresses, which is obtained from a triaxial compression test. We will do the same from now on and omit the subscripts, unless stated otherwise. In fact this is the friction angle that is used commonly in geotechnical and foundation designs.

The friction angle ϕ' of a granular soils increases with the angularity of the grains, surface roughness and relative density. Anecdotal evidence suggests some increase in friction angle with the grain size. Well graded granular soils generally have higher friction angle than the poorly graded ones. Wet soils have 1°–2° lower ϕ' than the dry soils. Figure 3.8 shows the friction angles determined from triaxial compression tests for different granular soils that have no plastic fines. Here, the friction angle is related to the soil type, relative density and unit weight. It can be seen that densely packed well graded gravels can have ϕ' as high as 45°. Even higher friction angles have been reported in the literature. Terzaghi and Peck (1967) suggested some representative values of ϕ' for sands and silts, as shown in Table 3.9. The relationship between the friction angle, the blow count from standard penetration test, and the relative density for sands is shown in Fig. 3.9.

3.3 Shear Strength Parameters c' and ϕ'

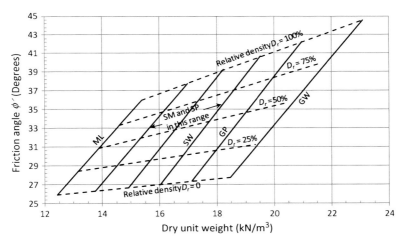

Fig. 3.8 Friction angles of granular soils (U.S. Navy 1982)

Table 3.9 Representative values of ϕ' for sands and silts

Soil	ϕ' (Degrees)	
	Loose	Dense
Sand, round grains, uniform	27.5	34
Sand, angular grains, well graded	33	45
Sandy gravels	35	50
Silty sand	27–33	30–34
Inorganic silt	27–30	30–35

After Terzaghi and Peck (1967)

Friction angle of a granular soil determined from triaxial compression tests ϕ'_{tc} was related to relative density by Schmertmann (1978) as shown in Fig. 3.10.

Some effective friction angle values suggested by the Australian Standard for earth retaining structures (AS 4678-2002) for soils and rocks are given in Table 3.10.

The peak effective friction angle ϕ'_{peak} of a granular soil can be written as (BS 8002 1994)

$$\phi'_{peak} = 30 + k_A + k_B + k_C \tag{3.28}$$

where, k_A, k_B and k_C account for the angularity of the grains (0°–4°), grain size distribution (0°–4°), and relative density expressed in terms of blow counts from the standard penetration test (0°–9°), respectively. These values are given in Table 3.11. The critical state friction angle ϕ'_{cv}, which is independent of the relative density, can be estimated as

	Very loose	Loose	Medium dense	Dense	Very dense
#D_r (%)	0　　15	35	65	85	100
*N_{60}	0　　4	10	30	50	
##$(N_1)_{60}$	0　　3(2)	8(5)	25(20)	42(35)	
**ϕ' (deg)	28	30	36	41	
##$(N_1)_{60}/D_r^2$		65	59	58	

*Terzaghi and Peck (1948); #Gibbs and Holtz (1957); ##Skempton (1986) with Tokimatsu and Seed (1987) in parentheses; **Peck et al. (1974)

Fig. 3.9 Relationship between relative density, friction angle and blow count from a standard penetration test for sands

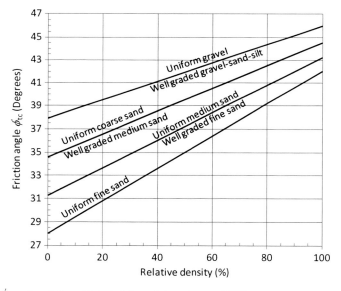

Fig. 3.10 $\phi'_{tc} - D_r$ relations (Adapted from Schmertmann 1978)

3.3 Shear Strength Parameters c' and ϕ'

Table 3.10 ϕ' for some soils and rocks as suggested by AS 4678-2002

	Material	$\phi'(°)$
Soils	Soft and firm clay of medium to high plasticity, silty clays, loose variable clayey fills, loose sandy silts (use $c' = 0$–5 kPa)	17–25
	Stiff sandy clays, gravelly clays, compacted clayey sands and sandy silts, compacted clay fill (use $c' = 0$–10 kPa)	26–32
	Gravelly sands, compacted sands, controlled crushed sandstone and gravel fills, dense well graded sands (use $c' = 0$–5 kPa)	32–37
	Weak weathered rock, controlled fills of roadbase, gravelly and recycled concrete (use $c' = 0$–25 kPa)	36–43
Rocks	Chalk	35
	Weathered granite	33
	Fresh basalt	37
	Weak sandstone	42
	Weak siltstone	35
	Weak mudstone	28

Table 3.11 k_A, k_B, k_C values for Eqs. (3.28) and (3.29)

		k – value
k_A	Rounded grains	0
	Sub-angular grains	2
	Angular grains	4
k_B	Uniformly graded ($C_u < 2$)	0
	Moderately graded ($2 < C_u < 6$)	2
	Well graded ($C_u > 6$)	4
k_C	$N_{60} < 10$	0
	$N_{60} = 20$	2
	$N_{60} = 40$	6
	$N_{60} = 60$	9

After AS 4678-2002

$$\phi'_{cv} = 30 + k_A + k_B \tag{3.29}$$

3.3.5 Stress Path Dependence of Friction Angles

In reality, the friction angle depends on the boundary conditions and the stress path followed to failure. In the field situations, within axisymmetric and plane strain conditions that are commonly assumed, there can be compressive or tensile loading. Some of the laboratory tests carried out to replicate the field situations are triaxial compression, triaxial extension, plane strain compression, plane strain extension, direct shear, direct simple shear, etc. Plane strain compression friction angle ϕ'_{psc} of a sand is 2°–7° greater than the direct shear friction angle ϕ'_{ds}. Allen et al. (2004) suggested that for granular soils

Table 3.12 Relative values of friction angles of cohesionless soils from different tests

Test type	Friction angle
Triaxial compression	$1.0\ \phi'_{tc}$
Triaxial extension	$1.12\ \phi'_{tc}$
Plane strain compression	$1.12\ \phi'_{tc}$
Plane strain extension	1.12 (for PSC to TC) × 1.12 (TE to TC) = $1.25\phi'_{tc}$
Direct shear test	$\tan^{-1}[\tan \phi'_{psc} \times \cos \phi'_{cv}] = \tan^{-1}[\tan 1.12\ \phi'_{tc} \times \cos \phi'_{cv}]$

Kulhawy and Mayne (1990)

Table 3.13 Relative values of friction angles of normally consolidated cohesive soils from different tests

Test type	Friction angle
Triaxial compression	$1.0\ \phi'_{tc}$
Triaxial extension	$1.22\ \phi'_{tc}$
Plane strain compression	$1.10\ \phi'_{tc}$
Plane strain extension	1.10 (for PSC to TC) × 1.22 (TE to TC) = $1.34\phi'_{tc}$
Direct shear test	$\tan^{-1}[\tan \phi'_{psc} \times \cos \phi'_{cv}] = \tan^{-1}[\tan 1.10\ \phi'_{tc} \times \cos \phi'_{cv}]$

Kulhawy and Mayne (1990)

$$\phi'_{psc} = \tan^{-1}(1.2 \tan \phi_{ds}) \qquad (3.30)$$

Direct shear friction angle ϕ'_{ds} of a sand can be greater or less than the triaxial compression friction angle ϕ'_{tc}, depending on ϕ'_{cv}, relative density and the stress level. The relationships among the values of ϕ' for cohesionless soils, as determined from the different tests are summarized in Table 3.12 (Kulhawy and Mayne 1990). Similar relationships for cohesive soils are given in Table 3.13.

It can be seen from Tables 3.12 and 3.13 that the conventional triaxial compression test gives the lowest possible values for ϕ'_{peak}. Therefore, using ϕ'_{tc} for other types of loadings, without any adjustment, can lead to conservative solutions. Castellanos and Brandon (2013) showed from an extensive database of tests conducted on riverine and lacustrine alluvial intact specimens from New Orleans, USA, that the effective friction angle from CU triaxial test is significantly greater than the ones from consolidated drained direct shear tests, and they both decrease with increasing PI (Fig. 3.11). Their ϕ' versus PI relationship can be approximated as:

$$\phi'_{tc(CU)} = 45 - \frac{PI}{0.5 + 0.04PI} \qquad (3.31)$$

and

$$\phi'_{ds} = 31 + 0.0017PI^2 - 0.3642PI \qquad (3.32)$$

3.3 Shear Strength Parameters c' and ϕ'

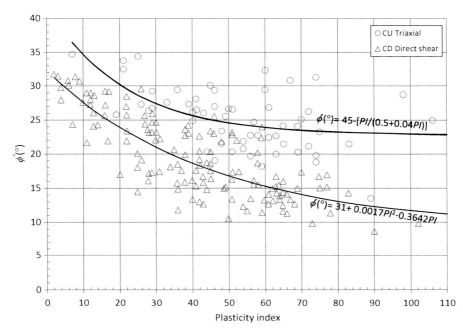

Fig. 3.11 ϕ' versus *PI* relationship for CU triaxial and direct shear tests on intact specimens (Adapted from Castellanos and Brandon 2013)

They also noted that the difference was insignificance in the case of remoulded clays. They attributed this to the destruction of the anisotropic fabric during remoulding which makes the shear strength independent of the failure plane orientation.

From the work of Bolton (1986), and supported by other test data, Schanz and Vermeer (1996) suggested that the peak friction angles of sands under triaxial and plane strain conditions are related by

$$\phi'_{tc} \approx \frac{1}{5}\left(3\phi'_{psc} + 2\phi'_{cv}\right) \tag{3.33}$$

They noted that while ϕ'_{peak} is significantly larger for plane strain compression than triaxial compression, the dilatancy angle ψ and the critical state friction angle ϕ'_{cv} appear to be the same for both loading conditions. It is also evident from Eq. (3.33) that the difference between ϕ'_{psc} and ϕ'_{tc} (both peak values) becomes smaller with lower relative densities where ϕ'_{tc} gets closer to ϕ'_{cv}.

In critical state soil mechanics, the slope of the failure envelope for triaxial compression loading in p'-q plane is M_c, given by

$$M_c = \frac{6\sin\phi'_{tc}}{3 - \sin\phi'_{tc}} \tag{3.34}$$

In triaxial extension, the slope M_e is given by

Table 3.14 Typical values of Skempton's A-parameter at failure

Soil	A_f
Sensitive clays	1.2–3
Normally consolidated clays	0.7–1.3
Overconsolidated clays	0.3–0.7
Heavily overconsolidated clays	−0.5–0
Very loose fine sand	2–3
Medium fine sand	0
Dense fine sand	−0.3
Loess	−0.2
Saturated silt, moderately dense	0.5

After Winterkorn and Fang (1975) and Leonards (1962)

$$M_e = \frac{6 \sin \phi'_{tc}}{3 + \sin \phi'_{tc}} \qquad (3.35)$$

3.3.6 Skempton's Pore Pressure Parameters

Skempton (1954) proposed a simple method to estimate the pore water pressure change in a saturated or partially saturated soil, when subjected to undrained loading under principal stress increments $\Delta\sigma_1$ and $\Delta\sigma_3$. The equation is given as

$$\Delta u = B[\Delta\sigma_3 + A(\Delta\sigma_1 - \Delta\sigma_3)] \qquad (3.36)$$

where A and B are known as Skempton's pore pressure parameters. B is a measure of the degree of saturation, which varies between 0 for dry soils and 1 for saturated soils. The A-parameter can vary during the shear, and is denoted by A_f at failure. Some typical values of A_f are given in Table 3.14.

From modified Cam clay model, it can be shown that A_f is given by

$$(A_f)_{CIUC} = 2^\Lambda + \frac{\frac{M}{3} - 1}{M} \qquad (3.37)$$

3.3.7 Sensitivity of Clays

The level of sensitivity observed in the clays varies geographically. Significantly greater values of sensitivity have been reported from Scandinavian countries, compared to those from Canada or USA. As a result, there are slightly different classification scales, which are shown in Table 3.15. High sensitivity is generally associated with high liquidity index. Scandinavian clays have liquidity index significantly larger than 1.0.

3.4 Undrained Shear Strength of a Clay c_u

Table 3.15 Sensitivity classification

Description	Sensitivity, S_t		
	U.S.	Canada[a]	Sweden
Low sensitive	2–4	1–2	<10
Medium sensitive	4–8	2–4	10–30
Highly sensitive	8–16	4–8	30–50
Extra sensitive	16	8–16	50–100
Quick	–	>16	>100

[a]Canadian Geotechnical Society (1992)

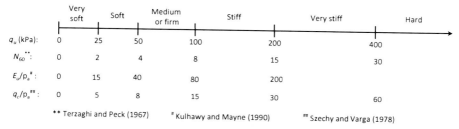

Fig. 3.12 Undrained strength classifications

3.4 Undrained Shear Strength of a Clay c_u

Cohesive soils can be classified based on the unconfined compressive strength q_u as shown in Fig. 3.12. Undrained shear strength c_u is given by half of q_u. For very stiff and hard clays, the water content would be less than the plastic limit. For very soft clays, the liquidity index is generally greater than 0.5. Although standard penetration test is not reliable in clays, when there are some data, they can be used in evaluating the consistency of the clay as shown in Fig. 3.12.

A rough estimate of the undrained shear strength can be obtained from (Hara et al. 1971; Kulhawy and Mayne 1990)

$$\frac{c_u}{p_a} = 0.29 \, N_{60}^{0.72} \qquad (3.38)$$

Equation (3.38) can give unrealistically high estimates of c_u. For K_0-consolidated soils, it can be shown from the first principles that

$$\left(\frac{c_u}{\sigma'_{vo}}\right)_{CK_0UC} = \frac{[K_0 + A_f(1 - K_0)]\sin\phi'_{tc}}{1 + (2A_f - 1)\sin\phi'_{tc}} \qquad (3.39)$$

For isotropically consolidated soils, the Eq. 3.39 reduces to

$$\left(\frac{c_u}{\sigma'_{vo}}\right)_{CIUC} = \frac{\sin \phi'_{tc}}{1 + (2A_f - 1)\sin \phi'_{tc}} \qquad (3.40)$$

The friction angle is the same for K_0 and isotropic consolidation (Mayne 1985; Kulhawy and Mayne 1990).

From modified Cam Clay model, it can be shown that for normally consolidated clays that are consolidated isotropically (Kulhawy and Mayne 1990).

$$\left(\frac{c_u}{\sigma'_{vo}}\right)_{CIUC} = 0.129 + 0.00435 PI \qquad (3.41)$$

From modified Cam clay model, it can also be shown that (Wroth and Houlsby 1985):

$$\left(\frac{c_u}{\sigma'_{vo}}\right)_{CIUC} = \frac{M}{2}\left(\frac{1}{2}\right)^{\Lambda} \qquad (3.42)$$

Wroth (1984) showed that

$$\frac{c_u}{\sigma'_{vo}} = \frac{\phi'_{cv}}{100} \qquad (3.43)$$

The undrained strength determined from isotropic consolidation and K_0 consolidation are related by (Kulhawy and Mayne 1990):

$$\left(\frac{c_u}{\sigma'_{vo}}\right)_{CK_0UC} = 0.15 + 0.49 \left(\frac{c_u}{\sigma'_{vo}}\right)_{CIUC} \qquad (3.44)$$

Equation (3.44) was obtained by regression analysis of 48 data points from different normally consolidated clays.

c_u/σ'_{vo} of normally consolidated clays in situ generally varies in the range of 0.2–0.3. Skempton (1957) suggested that for normally consolidated clays, based on vane shear test data,

$$\frac{c_u}{\sigma'_{vo}} = 0.0037 PI + 0.11 \qquad (3.45)$$

For overconsolidated, this ratio is larger and it increases with the overconsolidation ratio. Ladd et al. (1977) showed that

$$\left(\frac{c_u}{\sigma'_{vo}}\right)_{OC} = \left(\frac{c_u}{\sigma'_{vo}}\right)_{NC} OCR^{0.8} \qquad (3.46)$$

Jamiolkowski et al. (1985) suggested that for clays of low to moderate plasticity index

3.4 Undrained Shear Strength of a Clay c_u

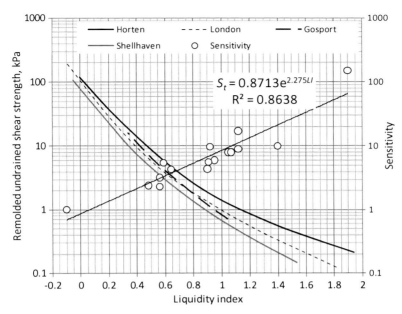

Fig. 3.13 Remolded undrained shear strength, liquidity index, and sensitivity relationship (Adapted from Skempton and Northey 1952)

$$\left(\frac{c_u}{\sigma'_{vo}}\right)_{OC} = (0.23 \pm 0.04)OCR^{0.8} \quad (3.47)$$

For overconsolidated clays of low to moderate plasticity, the above equation can also be approximated as (Jamiolkowski et al. 1985)

$$\left(\frac{c_u}{\sigma'_p}\right)_{OC} = 0.23 \pm 0.04 \quad (3.48)$$

Mesri (1989) suggested that $c_u/\sigma'_p = 0.22$ where σ'_p is the preconsolidation pressure.

In a triaxial compression test, the undrained shear strength increases with the increase in strain rate. A ten-fold (i.e. one log cycle) increase in the strain rate will increase the undrained shear strength by 10 %. Kulhawy and Mayne (1990) suggested strain rate $\dot{\varepsilon}$ of 1 % per hour as the standard reference rate, and the following equation to adjust the undrained shear strength to this reference rate.

$$[c_u]_{\dot{\varepsilon}=1\%/hour} = \frac{c_u}{1 + 0.1\log\dot{\varepsilon}} \quad (3.49)$$

Graham et al. (1983) reported that these trends are also true for direct simple shear tests and K_0 consolidated triaxial extension tests.

Skempton and Northey (1952) summarized some sensitivity – liquidity index data for some clays of moderate sensitivity, which are shown in Fig. 3.13.

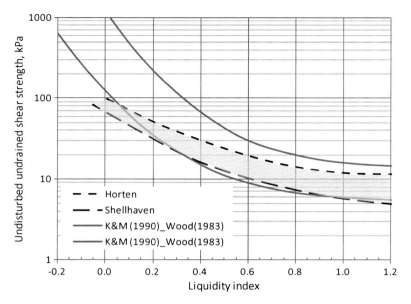

Fig. 3.14 Undisturbed undrained shear strength versus liquidity index derived from Fig. 3.13

The trend of sensitivity increasing with liquidity index is clear and they can be related by

$$S_t = 0.87 \exp(2.28 LI) \tag{3.50}$$

Skempton and Northey (1952) also produced the remolded undrained shear strength versus liquidity index variation for four different clays which fall into a narrow band in Fig. 3.13. At a specific *LI*, from the sensitivity estimated from Eq. (3.50) or Fig. 3.13, and the remolded undrained shear strength derived from the same figure, it is possible to estimate the undisturbed undrained shear strength. The shear strength values thus derived for undisturbed Horten and Shellhaven clays are used as the upper and lower bound of the shaded band shown in Fig. 3.14. Also shown in the figure is the band suggested by Wood (1983) and recommended by Kulhawy and Mayne (1990).

3.5 Soil Stiffness and Young's Modulus

Young's modulus E is the most common parameter used as a measure of stiffness. It is required in determining deformations, including settlements. In granular soils, it is derived through penetration test data from in situ tests such as standard

3.5 Soil Stiffness and Young's Modulus

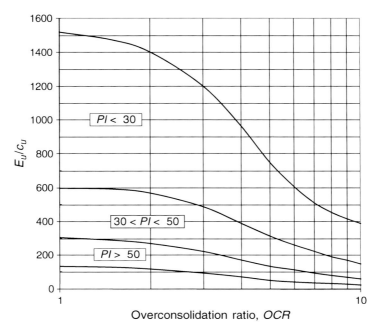

Fig. 3.15 $E_u/c_u - PI - OCR$ relation for clays

penetration test, cone penetration test, etc. Kulhawy and Mayne (1990) suggested that for normally consolidated clean sands

$$\frac{E}{p_a} \approx 10 N_{60} \qquad (3.51)$$

and for sands with fines,

$$\frac{E}{p_a} \approx 5 N_{60} \qquad (3.52)$$

where, N_{60} is the blow count from standard penetration corrected for energy rating. Schmertmann et al. (1978) suggested that for axisymmetric loading $E = 2.5\, q_c$ and for plane strain loading $E = 3.5\, q_c$, where q_c is the cone resistance from a cone penetration test.

The undrained modulus of clays (E_u) is generally estimated from an appropriate value of the modulus ratio E_u/c_u, which is generally in the range of 100–1000. It can be derived from Fig. 3.15 proposed by Duncan and Buchignani (1976) and the U.S. Army (1994). Typical values of E_u for different clay types, as recommended by U.S. Army (1994) are given in Table 3.16. Poisson's ratio ν of a material is defined as

Table 3.16 Typical values of E_u for clays

Clay	E_u (MPa)
Very soft clay	0.5–5
Soft clay	5–20
Medium clay	20–50
Stiff clay, silty clay	50–100
Sandy clay	25–200
Clay shale	100–200

After U.S. Army (1994) and Bowles (1986)

Table 3.17 Typical values of Poisson's ratio

Material	Poisson's ratio
Saturated clays (undrained)	0.5
Saturated clays (drained)	0.2–0.4
Dense sand	0.3–0.4
Loose sand	0.1–0.3
Loess	0.1–0.3
Ice	0.36
Aluminum	0.35
Steel	0.29
Concrete	0.15

Bowles (1986), Kulhawy and Mayne (1990), and Lambe and Whitman (1979)

$$\nu = -\frac{\text{lateral normal strain}}{\text{longitudinal normal strain}} \qquad (3.53)$$

and is in the range of 0–0.5 for most engineering materials. For saturated undrained clays, assuming no volume change, it can be shown theoretically that the Poisson's ratio is 0.5. Typical values of some soils are given in Table 3.17, along with other materials. Kulhawy and Mayne (1990) suggested that the drained Poisson's ratio of granular soils can be estimated as

$$\nu_d = 0.1 + 0.3 \frac{\phi'_{tc} - 25}{20} \qquad \text{for } \phi'_{tc} < 45° \qquad (3.54)$$

The drained Poisson's ratio of slightly overconsolidated clays increases slightly with *PI*, *OCR* and stress level.

Foundations on soils or rocks are designed to be safe against any possible bearing capacity failure and to undergo settlements that are within tolerable limits. Preliminary estimates of the footing dimensions can be arrived at on the basis of the presumed bearing capacity values given in Table 3.18. These values are generally conservative and should be used with caution.

Table 3.18 Presumed bearing capacity values

Group	Description	Presumed allowable bearing capacity (kPa)	Remarks
Coarse grained soil	Dense gravel or dense sand and gravel	>600	Width of footing $B > 1$ m. Water table at $> B$ below the footing.
	Compact gravel or compact sand and gravel	200–600	
	Loose gravel or loose sand and gravel	<200	
	Dense sand	>300	
	Compact sand	100–300	
	Loose sand	<100	
Fine grained soil	Very stiff or hard clays or heterogeneous mixtures such as tills	300–600	If $PI > 30$ and clay content >25 %, there are possible swell/shrink problems.
	Stiff clays	150–300	
	Firm clays	75–150	
	Soft clays and silts	<75	
	Very soft clays and silts	Not applicable	

After Canadian Geotechnical Society (1992)

3.6 Coefficient of Earth Pressure at Rest K_o

Treating the soil as a linear elastic continuum, it can be shown that

$$K_0 = \frac{\nu}{1 - \nu} \qquad (3.55)$$

where ν is the Poisson's ratio.

Jaky (1948) showed that for all particulate materials including soils

$$K_0 = \frac{1 - \sin \phi'}{1 + \sin \phi'} \left(1 + \frac{2}{3} \sin \phi'\right) \qquad (3.56)$$

Noting that $1 + \frac{\frac{2}{3}\sin\phi'}{1+\sin\phi'}$ is about 0.9 for typical values of friction angles, Jaky's original suggestion was to take K_0 as

$$(K_0)_{NC} = 0.9 \left(1 - \sin \phi'\right) \qquad (3.57)$$

which has been simplified and commonly used as

$$(K_0)_{NC} = 1 - \sin \phi' \qquad (3.58)$$

for normally consolidated soils. When the soils are overconsolidated, K_0 can be significantly greater, and can be estimated as (Brooker and Ireland 1965; European Committee for Standardization 1994)

$$(K_0)_{OC} = (K_0)_{NC}\sqrt{OCR} \tag{3.59}$$

Mayne and Kulhawy (1982) suggested that

$$(K_0)_{OC} = \left(1 - \sin\phi'\right) OCR^{\sin\phi'} \tag{3.60}$$

$(K_0)_{OC}$ is commonly expressed as

$$(K_0)_{OC} = (K_0)_{NC} OCR^m \tag{3.61}$$

where, m is an exponent commonly taken as 0.5, which is also suggested by the Eurocode (European Committee for Standardisation 1994). Ladd et al. (1977) suggested that $m = 0.42$ for low plastic clays and $m = 0.32$ for highly plastic clays.

For sloping ground, where σ'_v and σ'_h are no longer principal stresses, Kezdi (1972) extended Jaky's equation to

$$K_0 = \frac{1 - \sin\phi'}{1 + \sin\beta} \tag{3.62}$$

where β is the inclination of the slope to the horizontal. Brooker and Ireland (1965) showed that for normally consolidated clays

$$K_0 = 0.95 - \sin\phi' \tag{3.63}$$

Alpan (1967) showed that for normally consolidated clays

$$K_0 = 0.19 + 0.233 \log PI \tag{3.64}$$

Massarsch (1979) showed that for normally consolidated clays

$$K_0 = 0.44 + 0.0042 PI \tag{3.65}$$

Some typical values of K_0 reported in literature are summarized in Table 3.19 (Craig 2004).

Table 3.19 Typical values of K_0

Soil description	K_0
Dense sand	0.35
Loose sand	0.60
Normally consolidated clays (Norway)	0.5–0.6
Clay with OCR = 3.5 (London)	1.0
Clay with OCR = 20 (London)	2.8

Table 3.20 Relationships for the unit skin friction f_s in driven piles

Soil Type	Equation	Remarks	References
Clay	$f_s = \alpha c_u$	$\alpha = 1.0$ ($c_u \leq 25$ kN/m²) $\alpha = 0.5$ ($c_u \geq 70$ kN/m²) Linear variation in between	API (1984)
		$\alpha = 1.0$ ($c_u \leq 35$ kN/m²) $\alpha = 0.5$ ($c_u \geq 80$ kN/m²) Linear variation in between Length factor applies for $L/d > 50$	Semple & Rigden (1984)
		$\alpha = \left(\frac{c_u}{\sigma_v'}\right)_{nc}^{0.5} \left(\frac{c_u}{\sigma_v'}\right)^{-0.5}$ for $\left(\frac{c_u}{\sigma_{vo}'}\right) \leq 1$ $\alpha = \left(\frac{c_u}{\sigma_v'}\right)_{nc}^{0.5} \left(\frac{c_u}{\sigma_v'}\right)^{-0.25}$ for $\left(\frac{c_u}{\sigma_{vo}'}\right) \geq 1$	Fleming et al. (1985)
	$f_s = \beta \sigma_v$	$\beta = (1 - \sin \varphi') \tan \varphi' (OCR)^{0.5}$	Burland (1973) Meyerhof (1976)
Silica sand	$f_s = \beta \sigma_v'$ ($f_s \not> f_{slim}$)	$\beta = 0.15 - 0.35$ (compression) $0.10 - 0.24$ (tension)	McClelland (1974)
		$\beta = 0.44$ for $\varphi' = 28°$ 0.75 for $\varphi' = 35°$ 1.2 for $\varphi' = 37°$	Meyerhof (1976)
		$\beta = (K/K_o)K_o \tan(\varphi \cdot \delta/\varphi)$ δ/φ depends on interface materials (range 0.5–1.0); K/K_o depends on installation method (range 0.5–2.0) $K_o =$ coefficient of earth pressure at rest, and is a function of OCR	Stas and Kulhawy (1984)
Uncemented calcareous sand	$f_s = \beta \sigma_v'$	$\beta = 0.05 - 0.1$	Poulos (1988)

After Poulos (1989)

3.7 Using Laboratory Test Data in Pile Designs

The unit shaft resistance f_s for driven piles can be estimated as $f_s = \alpha c_u$ (total stress method) or $f_s = \beta \sigma_v'$ (effective stress method), where $\alpha =$ adhesion factor and $\beta = K_s \tan \delta$ with K_s and δ being the lateral earth pressure coefficient and the interfacial friction angle, respectively, at the soil-pile interface. The correlations for f_s of driven piles are summarized in Table 3.20. The unit skin friction correlations for the bored piles are given in Table 3.21. Correlations for the end bearing capacity of the pile tip f_b are given in Table 3.22.

Table 3.21 Relationships for the unit skin friction f_s in bored piles

Soil type	Equation	Remarks	References
Clay	$f_s = \alpha c_u$	$\alpha = 0.45$ (London clay)	Skempton (1959)
		$\alpha = 0.7$ times value for driven displacement pile	Fleming et al. (1985)
	$f_s = K \tan\delta \, \sigma_v'$	K is lesser of K_0 or $0.5(1+K_0)$	Fleming et al. (1985)
		$K/K_0 = 2/3$ to 1; K_0 is a function of OCR; δ depends on interface materials	Stas and Kulhawy (1984)
Silica sand	$f_s = \beta \sigma_v'$	$\beta = 0.1$ for $\varphi' = 33°$	Meyerhof (1976)
		$\beta = 0.2$ for $\varphi' = 35°$ $\beta = 0.35$ for $\varphi' = 37°$ $\beta = F \tan(\varphi - 5°)$ where $F = 0.7$ (compression) & 0.5 (tension)	Kraft and Lyons (1974)
Uncemented calcareous sand	$f_s = \beta \sigma_v'$ $(f_s \not> f_{slim})$	$\beta = 0.5$ to 0.8 $f_{slim} = 60$ to 100 kN/m^2	Poulos (1988)

After Poulos (1989)

Table 3.22 Relationships for the end bearing capacity

Soil type	Equation	Remarks	References
Clay	$f_b = N_c c_{ub}$	$N_c = 9$ for $L/D \geq 3$ c_{ub} = value of c_u in vicinity of pile tip	Skempton (1959)
Silica sand[1]	$f_b = N_q \sigma_v'$ $(f_s \not> f_{blim})^2$	$N_q = 40$	API (1984)
		N_q plotted against φ'	Berezantzev et al. (1961)
		N_q related to φ', relative density and mean effective stress	Felming et al. (1985)
		N_q from cavity expansion theory, as a function of φ' and volume compressibility	Vesic (1972)
Uncemented calcareous sand	$f_b = N_q \sigma_v'$ $(f_b \not> f_{blim})^2$	$N_q = 20$	Datta et al. (1980)
		Typical range of $N_q = 8$–20	Poulos (1988)
		N_q determined for reduced value of φ' (e.g. 18°)	Dutt and Ingram (1984)

After Poulos (1989)

Notes
1. For silica and calcareous sands, the above expressions apply for driven piles only
2. Typical limiting values f_{blim} range from 10 MN/m^2 to 15 MN/m^2 for silica sand, and 3–5 MN/m^2 for calcareous sand; the latter value depends on soil compressibility

References

Allen TM, Bathurst RJ, Holtz RD, Lee WF, Walters D (2004) New method for prediction of loads in steel reinforced soil walls. J Geotech Geoenviron Eng ASCE 130(11):1109–1120

Alpan I (1967) The empirical evaluation of the coefficient K_0 and K_{0R}. Soils Found 7(1):31–40

American Petroleum Institute (1984) Recommended practice for planning, designing and constructing fixed offshore platforms, *API RP2A*, 15th edn. American Petroleum Institute, Washington, DC

AS 4678-2002 Earth retaining structures, Australian Standard

Azzouz AS, Krizek RJ, Corotis RB (1976) Regression analysis of soil compressibility. Soils Found 16(2):19–29

Balasubramaniam AS, Brenner RP (1981) Chapter 7: Consolidation and settlement of soft clay. In: Brand EW, Brenner RP (eds) Soft clay engineering. Elsevier, Amsterdam, pp 481–566

Berezantzev VG, Khristoforov VS, Golubkov VN (1961) Load bearing capacity and deformation of piled foundations. In: Proceedings of 5th international conference on soil mechanics and foundation engineering, Paris, 2, pp 11–15

Bjerrum L, Simons NE (1960) Comparison of shear strength characteristics of normally consolidated clays. In: Proceedings of research conference on the shear strength of cohesive soils, ASCE, Boulder, Colorado, pp 711–726

Bolton MD (1986) The strength and dilatancy of sands. Geotechnique 36(1):65–78

Bowles JE (1988) Foundation analysis and design, 4th edn. McGraw-Hill, New York

Brooker EW, Ireland HO (1965) Earth pressures at rest related to stress history. Can Geotech J 2(1):1–15

BS 8002 (1994) Code of practice for earth retaining structures. British Standards Institution, London

Burland JB (1973) Shaft friction of piles in clay – a simple fundamental approach. Ground Eng 6(3):30–42

Burmister DM (1949) Principles and techniques of soil identification. In: Proceedings of annual highway research board meeting, National Research Council, Washington, DC, 29, pp 402–433

Canadian Geotechnical Society (1992) Canadian foundation engineering manual, 3rd edn, 511 pp

Carman PC (1938) The determination of the specific surfaces of powders. J Soc Chem Ind Trans 57:225

Carman PC (1956) Flow of gases through porous media. Butterworths Scientific Publications, London

Carrier WD III (2003) Good bye Hazen; Hello, Kozeny-Carman. J Geotech Geoenviron Eng ASCE 129(11):1054–1056

Carrier WD III, Beckman JF (1984) Correlations between index tests and the properties of remolded clays. Geotechnique 34(2):211–228

Castellanos BA,. Brandon TL (2013) A comparison between the shear strength measured with direct shear and triaxial devices on undisturbed and remolded soils. In: Proceedings of the 18th international conference on soil mechanics and geotechnical engineering, Paris, 1, pp 317–320

Chapuis RP (2004) Predicting the saturated hydraulic conductivity of sand and gravel using effective diameter and void ratio. Can Geotech J 41(5):787–795

Cozzolino VM (1961) Statistical forecasting of compression index. In: Proceedings of the 5th ICSMFE, Paris, pp 51–53

Craig RF (2004) Craig's soil mechanics, 7th edn. Spon Press/Taylor and Francis Group, London

Datta M, Gulhati SK, Rao GV (1980) An appraisal of the existing practise of determining the axial load capacity of deep penetration piles in calcareous sands. In: Proceedings of 12th annual OTC, Houston Paper OTC 3867, pp 119–130

Djoenaidi WJ (1985) A compendium of soil properties and correlations, MEngSc thesis, University of Sydney, Australia

Duncan JM, Buchignani AL (1976) An engineering manual for settlement studies, Geotechnical Engineering Report, Department of Civil Engineering, University of California, Berkeley, USA, 94 p

Dutt RN, Ingram WB (1984) Jackup rig siting in calcareous soils. In: Proceedings of 16th annual OTC, Houston Paper OTC 4840, 541–548

European Committee for Standardisation (1994) Eurocode 7: geotechnical design – Part 1, Brussels

Elnaggar MA, Krizek RJ (1970) Statistical approximation for consolidation settlement. Highway research record, no. 323, HRB, pp 87–96

Fleming WGK, Weltman AJ, Randolph MF, Elson WK (1985) Piling engineering. Surrey University Press/Wiley, Glasgow/New York

Goldberg GD, Lovell CW, Miles RD (1979) Use of geotechnical data bank, Transportation Research Record, No. 702, TRB, pp 140–146

Graham J, Crooks JHA, Bell AL (1983) Time effects on the stress-strain behavior of natural soft clays. Geotechnique 33(3):327–340

Hara A, Ohata T, Niwa M (1971) Shear modulus and shear strength of cohesive soils. Soils Found 14(3):1–12

Hazen A (1911) Discussion on "Dams on sand foundations". Trans ASCE 73:199

Hazen A (1930) "Water supply" American civil engineers handbook. Wiley, New York

Holtz RD, Kovacs WD (1981) An introduction to geotechnical engineering. Prentice-Hall, Englewood Cliffs

Hough BK (1957) Basic soils engineering. The Ronald Press Co., New York

Jaky J (1948) Pressures in silos. In: Proceedings of 2nd ICSMFE, Rotterdam, Holland, 1, pp 103–107

Jamiolkowski M, Ladd CC, Germaine JT, Lancellotta R (1985) New developments in field and laboratory testing of soils. In: Proceedings of the 11th international conference on soil mechanics and foundation engineering, San Francisco, 1, pp 57–154

Kenney TC (1959) Discussion of "Geotechnical properties of glacial lake clays," by T.H. Wu. J Soil Mech Found Div ASCE 85(SM3):67–79

Kezdi A (1972) Stability of rigid structures. In: Proceedings of 5th ECSMFE, Madrid, 2, pp 105–130

Koppula SD (1981) Statistical evaluation of compression index. Geotech Test J ASTM 4(2):68–73

Kozeny J (1927) ueber kapillareLeitung des Wassers im Boden, Wien, Akad Wiss 136(2a):271

Kraft LM, Lyons CG (1974) State-of-the art: ultimate axial capacity of grouted piles. In: Proceedings of 6th annual OTC, Houston, pp 487–503

Kulhawy FH, Mayne PW (1990) Manual on estimating soil properties for foundation design, Report EL-6800, Electric Power Research Institute, Palo Alto, California, USA

Ladd CC, Foott R, Ishihara K, Schlosser F, Poulos HG (1977) Stress-deformation and strength characteristics. In: Proceedings of 9th ICSMFE, Tokyo, 2, pp 421–494

Lambe TW, Whitman RV (1979) Soil mechanics SI version. Wiley, New York, 553 p

Leonards GA (1962) Foundation engineering. McGraw-Hill, New York

Lo YKT, Lovell CW (1982) Prediction of soil properties from simple indices, Transportation Research record, No. 873, Overconsolidated clays: Shales, TRB, pp 43–49

Massarsch KR (1979) Lateral earth pressure in normally consolidated clay. In: Proceedings of the 7th ECSMFE, Brighton, England, 2, pp 245–250

Mayne PW (1985) Stress anisotropy effects on clay strength. J Geotech Eng ASCE 111(3): 356–366

Mayne PW, Kulhawy FH (1982) K0-OCR relationships in soils. J Geotech Eng Div ASCE 108 (GT6):851–872

McClelland B (1974) Design of deep penetration piles for ocean structures. J Geotech Eng ASCE 100(GT 7):705–747

Mesri G (1973) Coefficient of secondary compression. J Soil Mech Found Div ASCE 99(SM1):123–137

Mesri G (1989) A reevaluation of $s_{u(mob)} \approx 0.22\ \sigma'_p$ using laboratory shear tests. Can Geotech J 26(1):162–164

Mesri G, Godlewski PM (1977) Time and stress compressibility interrelationship. J Geotech Eng Div ASCE 103(GT5):417–430

Mesri G, Olsen RE (1971) Mechanisms controlling the permeability of clays. Clay Clay Miner 19:151–158

Mesri G, Lo DOK Feng TW (1994) Settlement of embankments on soft clays. In: Proceedings of settlement '94, ASCE specialty conference, Geotechnical Special Publication No. 40, 1, pp 8–56

Meyerhof GG (1976) Bearing capacity and settlement of pile foundations. J Geotech Eng ASC 102(GT3):195–228

Mitchell JK (1976) Fundamentals of soil behavior. Wiley, New York

Nishida Y (1956) A brief note on compression index of soil. J Soil Mech Found Div, ASCE, 82 (SM3):1027-1 to 1027-14

Peck RB, Hanson WE, Thornburn TH (1974) Foundation engineering, 2nd edn. Wiley, New York

Poulos HG (1988) The mechanics of calcareous sediments. Jaeger Memorial Lecture, 5th Australia-New Zealand Geomechanics Conference, pp 8–41

Poulos HG (1989) Pile behavior – theory and application. Geotechnique 39(3):365–415

Rendon-Herrero O (1980) Universal compression index equation. J Geotech Eng Div ASCE 106(GT11):1179–1200

Salgado R (2008) The engineering of foundation. McGraw Hill, New York, 882 p

Schanz T, Vermeer PA (1996) Angles of friction and dilatancy of sand. Geotechnique 46(1):145–151

Schmertmann JH (1978) Guidelines for cone penetration test performance and design, Report FHWA-TS-78-209. U.S. Dept of Transportation, Washington, 145 pp

Schmertmann JH, Hartman JP, Brown PR (1978) Improved strain influence factor diagrams. J Geotech Eng Div ASCE 104(8):1131–1135

Semple RM, Rigden WJ (1984) Shaft capacity of driven piles in clay, Analysis and design of pile foundations, ASCE, pp 59–79

Skempton AW (1944) Notes on the compressibility of clays. Q J Geol Soc Lond 100:119–135

Skempton AW (1954) The pore pressure coefficients A and B. Geotechnique 4:143–147

Skempton AW (1957) Discussion on "The planning and design of the new Hong Kong airport". Proc Inst Civil Eng Lond 7:305–307

Skempton AW (1959) Cast-in-situ bored piles in London clay. Geotechnique 9(4):153–173

Skempton AW (1986) Standard penetration test procedures and the effects in sands of overburden pressure, relative density, particle size, ageing and overconsolidation. Geotechnique 36(3):425–447

Skempton AW, Northey RD (1952) The sensitivity of clays. Geotechnique 3:30–53

Sorensen KK, Okkels N (2013) Correlation between drained shear strength and plasticity index of undisturbed overconsolidated clays. In: Proceedings of the 18th international conference on soil mechanics and geotechnical engineering, Paris, Presses des Ponts, 1, pp 423–428

Sridharan A, Nagaraj HB (2000) Compressibility behavior of remoulded fine grained soils and correlations with index properties. Can Geotech J 37(3):712–722

Stas CV, Kulhawy FH (1984) Critical evaluation of design methods for foundations under axial uplift and compression loading, Report for EPRI, No. EL-3771, Cornell University

Szechy K, Varga L (1978) Foundation engineering – soil exploration and spread foundations. Akademiai Kiado, Budapest, 508 p

Taylor DW (1948) Fundamentals of soil mechanics. Wiley, New York, 700 pp

Terzaghi K, Peck R (1948) Soil mechanics in engineering practice. Wiley, New York

Terzaghi K, Peck R (1967) Soil mechanics in engineering practice, 2nd edn. Wiley, New York

Tezaghi K, Peck RB, Mesri G (1996) Soil mechanics in engineering practice, 3rd edn. Wiley, New York

Tokimatsu K, Seed HB (1987) Evaluation of settlements in sands due to earthquake shaking. J Geotech Eng ASCE 113(8):861–878

USACE (1990) Engineering and design – settlement analysis, EM1110-1-1904, Department of the Army, US Army Corps of Engineers

US Air Force (1983) Soils and geology procedures for foundation design of buildings and other structures, Air Force Manual AFM pp 88–3 Chapter 7 (Also U.S. Army Technical Manual 5-818-1), Departments of the Army and Air Force, Washington, DC

U.S.Army (1994) Settlement analysis, Technical Engineering and Design Guides, ASCE

U.S. Navy (1971) Soil mechanics, foundations and earth structures, NAVFAC Design manual DM-7, Washington, DC

U.S. Navy (1982) Soil mechanics – design manual 7.1, Department of the Navy, Naval Facilities Engineering Command, U.S. Government Printing Office, Washington, DC

Vesic AS (1972) Expansion of cavities in infinite soil mass. J Soil Mech Found Div ASCE 98:265–290

Winterkorn HF, Fang H-Y (1975) Foundation engineering handbook. Van Nostrand Reinhold Company, New York

Wood DM (1983) Index properties and critical state soil mechanics. In: Proceedings of the symposium on recent developments in laboratory and field tests and analysis of geotechnical problems, Asian Institute of Technology, Bangkok, pp 301–309

Wroth CP (1984) The interpretation of in situ soil tests. Geotechnique 34(4):449–489

Wroth CP, Houlsby GT (1985) Soil mechanics – property characterisation and analysis procedures. In: Proceedings of the 11th ICSMFE, San Francisco, pp 1–55

Wroth CP, Wood DM (1978) The correlation of index properties with some basic engineering properties of soils. Can Geotech J 15(2):137–145

Chapter 4
Standard Penetration Test

Abstract This chapter provides a detailed description of the Standard Penetration Test (SPT) procedure and corrections to be applied to the SPT N value and hammer energy. Correlations of SPT N value with relative density, peak drained friction angle and modulus of elasticity of sand are discussed in detail. In clays, correlations to obtain the undrained shear strength, preconsolidation pressure, over consolidation ratio are provided. As the SPT N value is used extensively in the design of foundations, correlations to obtain foundation bearing capacity for both shallow and deep foundations are provided.

Keywords Standard penetration test • SPT • Correlations • Relative density • Liquefaction • Bearing capacity

4.1 Standard Penetration Test Procedure

The standard penetration test (SPT) is a test conducted during a test boring in the field to measure the approximate soil resistance to penetration of a split-spoon sampler at various depths below the ground surface. The test allows a disturbed soil sample to be collected at various depths. This test is elaborated upon in ASTM Test Designation D-1586 (2014).

A section of a standard split-spoon sampler is shown in Fig. 4.1a. The tool consists of a steel driving shoe, a steel tube that is longitudinally split in half, and a coupling at the top. The coupling connects the sampler to the drill rod. The standard split tube has an inside diameter of 34.93 mm and an outside diameter of 50.8 mm. When a borehole is extended to a predetermined depth, a standard penetration test (SPT) can be conducted by removing the drill tools. The sampler is connected to the drill rod and lowered to the bottom of the hole. It is then driven into the soil by hammer blows to the top of the drill rod. The standard weight of the hammer is 622.72 N and, for each blow, the hammer drops a distance of 0.762 m. The number of blows required for a spoon penetration of three 152.4-mm intervals is recorded. The number of blows required for the last two intervals is added to give the standard penetration number N at that depth. This number is generally referred to as the N value. The sampler is then withdrawn, and the shoe and coupling are removed.

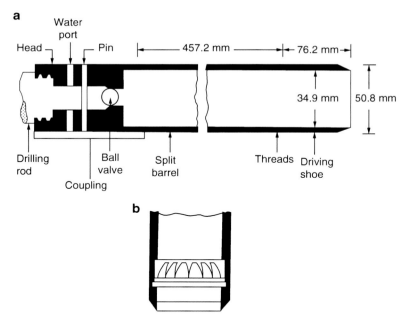

Fig. 4.1 (a) Standard split-spoon sampler; (b) spring core catcher

Finally, the soil sample recovered from the tube is placed in a glass bottle and transported to the laboratory.

The degree of disturbance for a soil sample is usually expressed as

$$A_R(\%) = \frac{D_o^2 - D_i^2}{D_i^2} \times 100 \qquad (4.1)$$

where

A_R = area ratio (ratio of disturbed area to total area of soil)
D_o = outside diameter of the sampling tube
D_i = inside diameter of the sampling tube

When the area ratio is 10 % or less, the sample generally is considered to be undisturbed. For a standard split spoon sampler,

$$A_R(\%) = \frac{(50.8)^2 - (34.93)^2}{(34.93)^2} \times 100 = 111.5\%$$

Hence, these samples are highly disturbed. Split-spoon samples generally are taken at intervals of about 1.5 m. When the material encountered in the field is sand (particularly fine sand below the water table), recovery of the sample by a split-

4.2 Correction of N Value for Effective Overburden Pressure (For Granular Soils)

spoon sampler may be difficult. In that case, a device such as a spring core catcher may have to be placed inside the split spoon (Fig. 4.1b).

In granular soils, the value of N is affected by the effective overburden pressure σ'_o. For that reason, the value of N obtained from field exploration under different effective overburden pressures should be changed to correspond to a standard value of σ'_o, or

$$N_1 = C_N N \quad (4.2)$$

where

N_1 = value of N corrected to a standard value of $\sigma'_o = p_a \approx 100 \text{ kN/m}^2$
C_N = correction factor
p_a = atmospheric pressure $\approx 100 \text{ kN/m}^2$
σ'_o = effective overburden pressure

A number of empirical relations have been proposed for the correction factor, C_N, in the past. Some of the relationships are given next. The most commonly cited relationships are those of Liao and Whitman (1986) and Skempton (1986).

- Liao and Whitman (1986):

$$C_N = \left[\frac{1}{\left(\frac{\sigma'_o}{p_a}\right)}\right]^{0.5} \quad (4.3)$$

- Skempton (1986):

$$C_N = \frac{2}{1 + \left(\frac{\sigma'_o}{p_a}\right)} \quad \text{(for normally consolidated fine sand)} \quad (4.4)$$

$$C_N = \frac{3}{2 + \left(\frac{\sigma'_o}{p_a}\right)} \quad \text{(for normally consolidated coarse sand)} \quad (4.5)$$

$$C_N = \frac{1.7}{0.7 + \left(\frac{\sigma'_o}{p_a}\right)} \quad \text{(for overconsolidated sand)} \quad (4.6)$$

- Tokimatsu and Yoshimi (1983):

$$C_N = \frac{1.7}{0.7 + \left(\frac{\sigma'_o}{p_a}\right)} \tag{4.7}$$

- Seed et al. (1975):

$$C_N = 1 - 1.25 \log\left(\frac{\sigma'_o}{p_a}\right) \tag{4.8}$$

- Peck et al. (1974):

$$C_N = 0.77 \log\left[\frac{20}{\left(\frac{\sigma'_o}{p_a}\right)}\right] \quad \left(\text{for } \frac{\sigma'_o}{p_a} \geq 0.25\right) \tag{4.9}$$

- Bazaraa (1967):

$$C_N = \frac{4}{1 + 4\left(\frac{\sigma'_o}{p_a}\right)} \quad \left(\text{for } \frac{\sigma'_o}{p_a} \leq 0.75\right) \tag{4.10}$$

$$C_N = \frac{4}{3.25 + \left(\frac{\sigma'_o}{p_a}\right)} \quad \left(\text{for } \frac{\sigma'_o}{p_a} > 0.75\right) \tag{4.11}$$

Table 4.1 shows the comparison of C_N derived using various relationships cited above [Eqs. (4.3), (4.4), (4.5), (4.6), (4.7), (4.8), (4.9), (4.10), and (4.11)]. It can be seen that the magnitude of the correction factor estimated by using any one of the relationships is approximately the same, considering the uncertainties involved in conducting the standard penetration tests. Hence, it appears that Eq. (4.3) may be used for all calculations.

Salgado et al. (1997, 2008) have suggested that Eq. (4.3) does not consider the lateral stress effect in soil. For the lateral stress effect to be accounted for, Eq. (4.3) may be modified as,

$$C_N = \left[\frac{1}{\left(\frac{\sigma'_o}{p_a}\right)} \times \frac{K_{o,\text{NC}}}{K_o}\right]^{0.5} \tag{4.12}$$

where (for granular soils)

$$K_{o,\text{NC}} = \text{at-rest earth pressure for normally consolidated soil}$$
$$= 1 - \sin\phi \tag{4.13}$$

ϕ = soil friction angle

4.3 Correction for SPT Hammer Energy Efficiency

Table 4.1 Variation of C_N

$\frac{\sigma'_o}{p_a}$	C_N						
	Eq. (4.3)	Eq. (4.4)	Eq. (4.5)	Eqs. (4.6) and (4.7)	Eq. (4.8)	Eq. (4.9)	Eqs. (4.10) and (4.11)
0.25	2.00	1.60	1.33	1.78	1.75	1.47	2.00
0.50	1.41	1.33	1.20	1.17	1.38	1.23	1.33
0.75	1.15	1.14	1.09	1.17	1.15	1.10	1.00
1.00	1.00	1.00	1.00	1.00	1.00	1.00	0.94
1.50	0.82	0.80	0.86	0.77	0.78	0.87	0.84
2.00	0.71	0.67	0.75	0.63	0.62	0.77	0.76
3.00	0.56	0.50	0.60	0.46	0.40	0.63	0.65
4.00	0.50	0.40	0.60	0.36	0.25	0.54	0.55

K_o = at-rest earth pressure for overconsolidated soil

$$K_o = (1 - \sin\phi)\sqrt{OCR} \qquad (4.14)$$

where

OCR = overconsolidation ratio = σ'_o/σ'_c
σ'_c = effective preconsolidation pressure

4.3 Correction for SPT Hammer Energy Efficiency

There are several factors that contribute to the variation of the standard penetration number N at a given depth for similar soil profiles. Among these factors are the SPT hammer efficiency, borehole diameter, sampling method, and rod length (Skempton 1986; Seed et al. 1985). The SPT hammer efficiency can be expressed as

$$E_r(\%) = \frac{\text{actual hammer energy to the sampler}}{\text{theoretical input energy}} \times 100 \qquad (4.15)$$

$$\text{Theoretical input energy} = Wh \qquad (4.16)$$

where

W = weight of the hammer ≈ 0.623 kN
h = height of drop ≈ 0.76 mm

So,

$$Wh = (0.623)(0.76) = 0.474 \text{ kN-m}$$

In the field, the magnitude of E_r can vary from 40 to 90 %. The standard practice now in the U.S. is to express the N value to an average energy ratio of 60 % ($\approx N_{60}$).

Table 4.2 Variation of η_H [Eq. (4.17)]

Country	Hammer type	Hammer release	η_H (%)
Japan	Donut	Free fall	78
	Donut	Rope and pulley	67
United States	Safety	Rope and pulley	60
	Donut	Rope and pulley	45
Argentina	Donut	Rope and pulley	45
China	Donut	Free fall	60
	Donut	Rope and pulley	50

Table 4.3 Variation of η_B [Eq. (4.17)]

Diameter (mm)	η_B
60–120	1
150	1.05
200	1.15

Table 4.4 Variation of η_S [Eq. (4.17)]

Variable	η_S
Standard sampler	1.0
With liner for dense sand and clay	0.8
With liner for loose sand	0.9

Thus correcting for field procedures, and on the basis of field observations, it appears reasonable to standardize the field penetration number as a function of the input driving energy and its dissipation around the sampler into the surrounding soil, or

$$N_{60} = \frac{N\eta_H\eta_B\eta_S\eta_R}{60} \qquad (4.17)$$

where

N_{60} = standard penetration number, corrected for field conditions
N = measured penetration number
η_H = hammer efficiency (%)
η_B = correction for borehole diameter
η_S = sampler correction
η_R = correction for rod length

Variations of η_H, η_B, η_S, and η_R, based on recommendations by Seed et al. (1985) and Skempton (1986), are summarized in Tables 4.2, 4.3, 4.4, and 4.5.

The typical value of E_r in the United States is about 55–70 %.

The N_{60} value can be corrected to a standard value of $\sigma_o' = p_a \approx 100$ kN/m² as,

$$(N_1)_{60} = C_N N_{60} \qquad (4.18)$$

where

C_N = correction factor given in Eqs. (4.3), (4.4), (4.5), (4.6), (4.7), (4.8), (4.9), (4.10), and (4.11)

4.4 Correlation of Standard Penetration Number with Relative Density (D_r) of Sand

Table 4.5 Variation of η_R [Eq. (4.17)]

Rod length (m)	η_R
>10	1.0
6–10	0.95
4–6	0.85
0–4	0.75

Hence, it is possible that, depending upon the source, one will be working with four different standard penetration numbers in various correlations available in the literature. They are:

1. N – penetration number obtained from the field
2. N_{60} – field N value corrected to an average energy ratio of 60 %
3. N_1 – N value obtained from field corrected to a standard effective overburden pressure $\sigma'_o = p_a \approx 100 \text{ kN/m}^2$
4. $(N_1)_{60}$ – N_{60} corrected to a standard effective overburden pressure $\sigma'_o = p_a \approx 100 \text{ kN/m}^2$

The N value for a given average energy ratio can be approximately converted to an N value for a different energy as follows:

$$N_{ER(1)} \times ER_{(1)} = N_{ER(2)} \times ER_{(2)}$$

where

$N_{ER(1)} = N$ value for an energy ratio $ER_{(1)}$
$N_{ER(2)} = N$ value for an energy ratio $ER_{(2)}$

For example, if $ER_{(1)} = 60$ % and $N_{ER(1)} = 12$ then, for $ER_{(2)} = 75$ %

$$N_{ER(2)} = \frac{N_{ER(1)} \times ER_{(1)}}{ER_{(2)}} = \frac{12 \times 60}{75} = 9.6 \approx 10$$

4.4 Correlation of Standard Penetration Number with Relative Density (D_r) of Sand

Terzaghi and Peck (1967) gave a qualitative description of relative density of sand, D_r, based on standard penetration number, N, which is given in Table 4.6.

Based on the early research in calibration chamber tests of Gibbs and Holtz (1957) provided relations for STP N-value, σ'_o/p_a, and relative density D_r. These are shown in Fig. 4.2. At a later stage, Holtz and Gibbs (1979) presented the correlation of N and D_r in a more usable form. This is shown in Fig. 4.3. Further research has shown that the relationships is somewhat more complex.

Table 4.6 Qualitative description of relative density (Based on Terzaghi and Peck 1967)

Standard penetration number, N	Relative density, D_r
0–4	Very loose
4–10	Loose
10–30	Medium
30–50	Dense
Over 50	Very dense

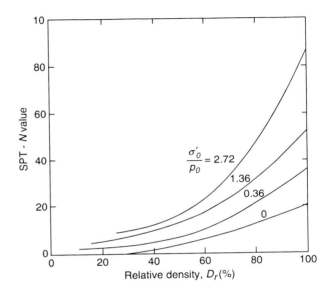

Fig. 4.2 Variation of N with σ'_o/p_a and D_r, (Adapted after Gibbs and Holtz 1957)

Meyehof (1957) provided a correlation between D_r and N in the form

$$N = \left(17 + 24\frac{\sigma'_o}{p_a}\right)D_r^2 \tag{4.19}$$

Or

$$D_r(\%) = 20.4\left(\frac{N}{0.7 + \frac{\sigma'_o}{p_a}}\right)^{0.5} \tag{4.20}$$

The standard penetration number given in Eqs. (4.19) and (4.20) are approximately equal to N_{60}. It gives fairly good estimates for clean, medium fine sands.

Kulhawy and Mayne (1990) analyzed several unaged normally consolidated sand and provided the following correlation for relative density, D_r, of sand:

4.4 Correlation of Standard Penetration Number with Relative Density (D_r) of Sand

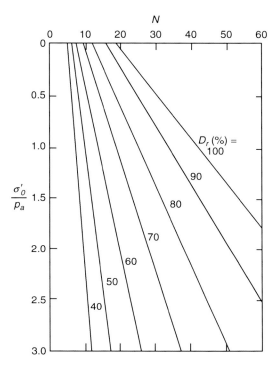

Fig. 4.3 Variation of D_r with N and σ'_o/p_a (Adapted after Holtz and Gibbs 1979)

$$\frac{(N_1)_{60}}{D_r^2} = 60 + 25\log D_{50} \tag{4.21}$$

Or

$$D_r(\%) = \left[\frac{(N_1)_{60}}{60 + 25\log D_{50}}\right]^{0.5} \times 100 \tag{4.22}$$

where

D_{50} = median grain size (mm)

Kulhawy and Mayne (1990) provided a modification of Eq. (4.22) to account for the effect of aging and overconsolidation in the form,

$$D_r(\%) = \left[\frac{(N_1)_{60}}{(60 + 25\log D_{50})C_A C_{OCR}}\right]^{0.5} \tag{4.23}$$

where

C_A = correction factor for aging
C_{OCR} = correction for overconsolidation

The correlations for C_A and C_{OCR} can be expressed as,

$$C_A = 1.2 + 0.05\log\left(\frac{t}{100}\right) \tag{4.24}$$

$t =$ time, in years, since deposition

$$C_{OCR} = (OCR)^{0.18} \tag{4.25}$$

$OCR =$ overconsolidation ratio

Marcuson and Bieganousky (1977) provided a correlation for D_r as

$$D_r(\%) = 12.2 + 0.75\left[222N + 2311 - 711OCR - 779\left(\frac{\sigma_o'}{p_a}\right) - 50C_u^2\right]^{0.5} \tag{4.26}$$

where
$C_u =$ uniformity coefficient

$$C_u = \frac{D_{60}}{D_{10}} \tag{4.27}$$

$D_{60}, D_{10} =$ diameter through which, respectively, 60 % and 10 % of the soil passes through Eq. (4.26) clearly shows that the grain-size distribution is another factor in the correlations for the relative density of sand.

Cubrinovski and Ishihara (1999, 2002) provided the results of an elaborate study related to maximum (e_{max}) and minimum (e_{min}) void ratios, median grain size (D_{50}), and standard penetration number (N) of sand and gravel. The experimental evidence from these results shows that, for granular soils (clean sand, sand with fines, and gravel), the difference between the maximum and minimum void ratios can be related to the median grain size as (Fig. 4.4),

$$e_{max} - e_{min} = 0.23 + \frac{0.06}{D_{50}} \tag{4.28}$$

where

$e_{max} =$ maximum void ratio
$e_{min} =$ minimum void ratio
$D_{50} =$ median grain size

Tests on high-quality undisturbed samples of silty sand, clean sand, and gravel deposits provide the following correlations (Fig. 4.5)

4.4 Correlation of Standard Penetration Number with Relative Density (D_r) of Sand   97

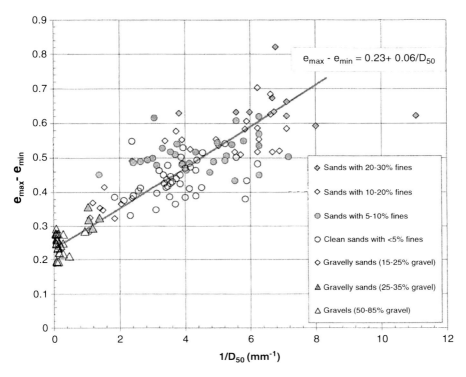

Fig. 4.4 Plot of e_{max}-e_{min} vs. $1/D_{50}$ (mm-1) – Test results of Cubrinovski and Ishihara (1999, 2002) for sands and gravels

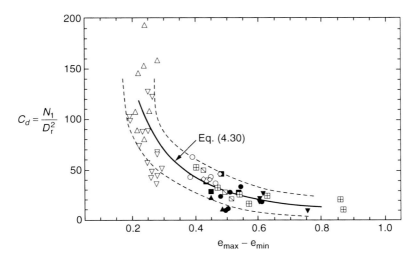

Fig. 4.5 Development of Eq. (4.30) (Adapted after Cubrinovski and Ishihara 1999)

$$C_D = \frac{N_1}{D_r^2} \qquad (4.29)$$

$$C_D = \frac{9}{(e_{max} - e_{min})^{1.7}} \qquad (4.30)$$

Hence,

$$\frac{N_1}{D_r^2} = \frac{9}{(e_{max} - e_{min})^{1.7}}$$

Or

$$D_r = \left[\frac{N_1(e_{max} - e_{min})^{1.7}}{9}\right]^{0.5} \qquad (4.31)$$

From Eqs. (4.2) and (4.3)

$$N_1 = N\left(\frac{\sigma_o'}{p_a}\right)^{0.5} \qquad (4.32)$$

Combining Eqs. (4.31) and (4.32),

$$D_r(\%) = \left[\frac{N(e_{max} - e_{min})^{1.7}}{9}\left(\frac{p_a}{\sigma_o'}\right)^{0.5}\right]^{0.5} \times 100 \qquad (4.33)$$

The N value reported in Eq. (4.33) approximately relates average energy ratio of 78 % (see Sect. 4.3), or

$$N \approx N_{78} \qquad (4.34)$$

Figure 4.6 shows a comparison of $D_{r\text{-measured}}$ with $D_{r\text{-calculated}}$ via Eq. (4.32).
Yoshida et al. (1988) suggested the following equation to estimate D_r, or

$$D_r(\%) = C_0\left(\sigma_o'\right)^{-C_1} N^{C_2} \qquad (4.35)$$

The unit of σ_o' is kN/m^2.

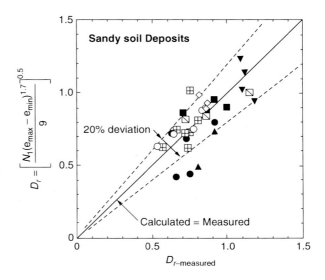

Fig. 4.6 Comparison of $D_{r\text{-}measured}$ with $D_{r\text{-}calculated}$ from Eq. (4.31) for sandy soil deposits (Adapted after Cubrinovski and Ishihara 1999)

The range of values of C_0, C_1, and C_2 is as follows:

Parameter	Range	Best-fit value
C_0	18–25	25
C_1	0.12–0.14	0.12
C_2	0.44–0.57	0.46

For practical purposes, $N \approx N_{60}$ (Bowles 1996). Hence, with the best-fit values,

$$D_r(\%) = 25\left(\sigma_o' \text{ kN/m}^2\right)^{-0.12} N_{60}^{0.46} \qquad (4.36)$$

4.5 Correlation of N with Peak Drained Friction Angle (ϕ) for Sand

There are several correlations available in the literature for the approximate value of peak drained friction (triaxial) angle (ϕ) with the standard penetration number. One of the early correlations was suggested by Meyerhof (1959) in the form

$$\phi = 28 + 0.15 D_r \qquad (4.37)$$

where

D_r = relative density, in per cent.

The approximate value of D_r can be obtained from the relations given in Sect. 4.4.

Fig. 4.7 Variation of ϕ with N and σ'_o/p_a (Adapted after DeMello 1971)

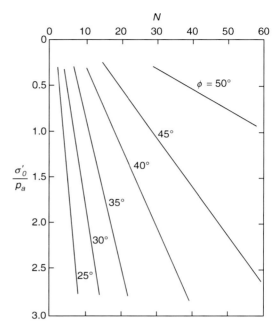

DeMello (1971) provided a correlation between N and ϕ which was based on the work of Gibbs and Holtz (1957) in a calibration chamber. This correlation is shown in Fig. 4.7. The N value shown in the figure is approximately equal to N_{60}.

Peck, Hanson, and Thornburn (1974) provided a correlation between N and ϕ as shown in Fig. 4.8. This correlation is probably a conservative one. Wolff (1989) approximated the relationship shown in Fig. 4.8 as

$$\phi(\deg) = 27.1 + 0.3N_1 - 0.00054N_1^2 \quad (4.38)$$

Schmertmann (1975) proposed a correlation between effective overburden pressure σ'_o, N, and ϕ, and this correlation is shown in Fig. 4.9. Kulhawy and Mayne (1990) approximated this correlation in the form

$$\phi = \tan^{-1}\left[\frac{N}{12.2 + 20.3\left(\frac{\sigma'_o}{p_a}\right)}\right]^{0.34} \quad (4.39)$$

Shioi and Fukui (1982) gave correlations between N and ϕ (obtained from the Japanese Railway Standards) which was slightly modified by Bowles (1996) and follows:

Fig. 4.8 Variation of soil friction angle – Eq. (4.37)

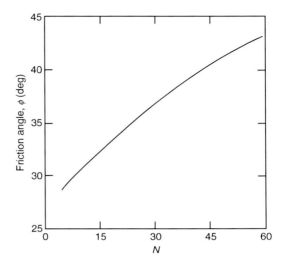

Fig. 4.9 Variation of ϕ with N and σ'_o/p_a, (Adapted after Schmertmann 1975)

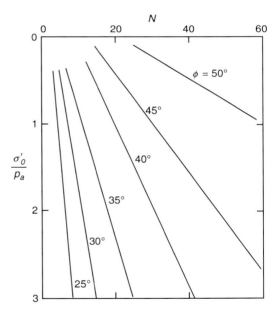

$$\phi = \sqrt{18(N_1)_{70}} + 15 \quad \text{(for roads and bridges)} \quad (4.40)$$

$$\phi = 0.36 N_{70} + 27 \quad \text{(for buildings)} \quad (4.41)$$

where
N_{70} = standard penetration number for 70 % energy ratio

Hatanaka and Uchida (1996) expressed the soil friction angle and corrected standard penetration number in the following form

$$\phi = \sqrt{20(N_1)_{60}} + 20 \tag{4.42}$$

4.6 Correlation of N with Modulus of Elasticity (E) for Sandy Soils

Several empirical relationships for the modulus of elasticity have been presented in the past and were summarized by Mitchell and Gardner (1975). However, those correlations show considerable scatter since the N value depends on several factors that are not clear from the original studies. Following are a few of those correlations:

- Webb (1969)
  For sand and clayey sand,

$$E = 5(N + 15) \text{ ton/ft}^2 \tag{4.43}$$

In SI units,

$$E \approx 479(N + 15) \text{ kN/m}^2 \tag{4.44}$$

- Ferrent (1963)

$$E = 7.5(1 - v^2)N \text{ ton/ft}^2 \tag{4.45}$$

$$v = \text{Poisson's ratio}$$

In SI units,

$$E \approx 718(1 - v^2)N \text{ kN/m}^2 \tag{4.46}$$

- Begemann (1974)
  For silt with sand to gravel with sand,

$$E = 40 + C(N - 6) \text{ kg/cm}^2 \quad (\text{for } N > 15) \tag{4.47a}$$

or,

$$E \approx 4000 + 100C(N - 6) \text{ kN/m}^2 \tag{4.47b}$$

and

$$E = C(N + 6) \text{ kg/cm}^2 (\text{for } N < 15) \tag{4.48}$$

or,

$$E \approx 100C(N+6) \text{ kN/m}^2 \quad (4.49)$$

$C = 3$ for silt with sand to 12 for gravel with sand

Due to the uncertainty of estimating E, Kulhawy and Mayne (1990) proposed the following as an initial approximation.

$$\frac{E}{p_a} = 5N_{60} \quad \text{(for sand with fines)} \quad (4.50)$$

$$\frac{E}{p_a} = 10N_{60} \quad \text{(for clean normally consolidated sand)} \quad (4.51)$$

$$\frac{E}{p_a} = 15N_{60} \quad \text{(for clean overconsolidated sand)} \quad (4.52)$$

4.7 Correlation of Undrained Cohesion (c_u) with N for Clay Soil

Table 4.7 provides the approximate consistency, corresponding N value, and undrained cohesion (c_u) of clay soils (Terzaghi and Peck 1967). These values need to be used with care.

From the table it is evident that

$$c_u(\text{kN/m}^2) \approx KN \quad (4.53)$$

where $K \approx 6$ and $N \approx N_{60}$

Szechy and Varga (1978) provided a correlation between the consistency index (CI), N, and c_u. The consistency index is defined as,

$$CI = \frac{LL - w}{LL - PL} \quad (4.54)$$

where

LL = liquid limit
PL = plastic limit
w = natural moisture content

Table 4.7 Approximate variation of consistency, N, and undrained cohesion of clay

Consistency	N	c_u (kN/m^2)
Very soft	0–2	<12
Soft	2–4	12–25
Medium	4–8	25–50
Stiff	8–15	50–100
Very stiff	15–30	100–200
Hard	>30	>200

Table 4.8 Variation of CI, N, and c_u

N	CI	Consistency	c_u (kN/m^2)
<2	<0.5	Very soft	<12.5
2–8	0.5–0.75	Soft to medium	12.5–40
8–15	0.75–1.0	Stiff	40–75
15–30	1.0–1.5	Very stiff	75–200
>30	>1.5	Hard	>200

Based on Szechy and Varga (1978)

Table 4.9 Variation of α' with plasticity index (PI)

PI	α'
15	0.068
20	0.055
25	0.048
30	0.045
40	0.044
60	0.043

The correlation is given in Table 4.8

Stroud (1975) provided a correlation between N, c_u, and plasticity index (PI) of clay soils, which is of the form

$$\frac{c_u}{p_a} = \alpha N \qquad (4.55)$$

The hammer used for obtaining the data had an energy ratio of approximately 73 %. For an energy ratio of 60 %, the α value provided by Stroud (1975) have been modified by Salgado (2008) as,

$$\alpha' = \frac{73}{60}\alpha \approx 1.22\alpha$$

Hence, Eq. (4.54) can be rewritten as

$$\frac{c_u}{p_a} = \alpha' N_{60} \qquad (4.56)$$

The interpolated values of α' with plasticity index are given in Table 4.9.

4.8 Correlation of Preconsolidation Pressure (σ'_c) with N for Clay Soil

Mayne and Kemper (1988) analyzed standard penetration test results of 50 different clay deposits along with the results of oedometer tests performed on thin-wall tube specimens. The correlation of those tests showed that

$$\frac{\sigma'_c}{p_a} = 67N^{0.83} \tag{4.57}$$

where σ'_c = preconsolidation pressure

Equation (4.57) can also be approximated as

$$\frac{\sigma'_c}{p_a} \approx 48N \tag{4.58}$$

It is important to point out that the N values [based on which Eqs. (4.57) and (4.58) have been developed] were not corrected to a standard energy ratio (i.e., N_{60}). So this may be considered as a first approximation only.

4.9 Correlation of Overconsolidation Ratio (*Ocr*) with N for Clay Soil

Mayne and Kemper (1988) provided regression analysis of 110 data points to obtain a correlation between N and overconsolidation ratio (*OCR*) of clay soil. According to this analysis

$$OCR = 0.193 \left(\frac{N}{\sigma'_o} \right)^{0.689} \tag{4.59}$$

where σ'_o is in MN/m².

For a forced exponent = 1, the regression data indicates

$$OCR = 0.193 \left(\frac{N}{\sigma'_o} \right) \tag{4.60}$$

4.10 Correlation of Cone Penetration Resistance (q_c) with N

Geotechnical engineers do not always have the luxury of having the standard penetration test data and the cone penetration test data. When only one is available, it is useful to have some means of converting from one to the other. Section 5.9 in Chap. 5 provides a detailed discussion and available correlations for q_c and N.

4.11 Correlation of Liquefaction Potential of Sand with N

During earthquakes, major destruction of various types of structures occurs due to the creation of fissures, abnormal and/or unequal movement, and loss of strength or stiffness of the ground. The loss of strength or stiffness of the ground results in the settlement of buildings, failure of earth dams, landslides, and other hazards. The process by which loss of strength occurs in soil is called *soil liquefaction*. The phenomenon of soil liquefaction is primarily associated with medium- to fine-grained *saturated cohesionless soils*. Example of soil liquefaction-related damage are the June 16, 1964 earthquake at Niigata, Japan, and also the 1964 Alaskan earthquake. Since the mid-1960s intensive investigations have been carried out around the world to determine the soil parameters that control liquefaction.

After the occurrence of the Niigata earthquake of 1964, Kishida (1966), Kuizumi (1966), and Ohasaki (1966) studied the areas in Niigata where liquefaction had and had not occurred. They developed criteria, based primarily on the standard penetration number of sand deposits, to differentiate between liquefiable and non-liquefiable conditions.

Following that, Seed (1979) used the results of several studies to develop the lower-bound correlation curves between the cyclic stress ratio in the field $(\tau_h/\sigma'_o)_{field}$ and $(N_1)_{60}$ for earthquake magitudes (M) of 6, 7.5, and 8.25. This is shown in Fig. 4.10, in which τ_h = peak cyclic shear stress and σ'_o = initial effective overburden pressure.

For given values of $(N_1)_{60}$ and M, if $(\tau_h/\sigma'_o)_{field}$ falls above the plot, then liquefaction is likely to occur. For estimation of τ_h, the readers may refer to Das and Ramana (2011).

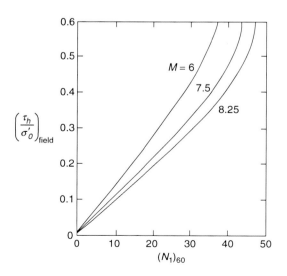

Fig. 4.10 Lower bound relationship for liquefaction-variation of $(\tau_h/\sigma'_o)_{field}$ with $(N_1)_{60}$ and M (Adapted after Seed 1979)

Semi-empirical field based procedures have been given strong attention in the last 2–3 decades and SPT, CPT (cone penetrometer test) and shear wave velocity have become the main in situ tests favoured by academics and practitioners to assess liquefaction potential. Although SPT has been the main test historically used, the CPT is becoming more common in liquefaction assessments (Robertson and Wride 1998) especially as the database of case histories grows. Semi-empirical procedures for evaluating liquefaction potential during earthquakes is well established and in 1996 NCEER workshop achieved a worldwide consensus on the semi-empirical assessment prevailing at the time. The procedures have been further developed as more data has become available and more research has been carried out. Section 5.11 in Chap. 5 discusses one of the methodologies currently adopted to assess liquefaction potential for both SPT and CPT tests.

4.12 Correlations for Shear Wave Velocity, v_s

Several correlations between the shear wave velocity v_s and field standard penetration number N have been presented in the past. A few of these correlations are given in Table 4.10. Significant differences exist among the published relations that may be due to differences in geology along with the measurement of N and v_s. If shear wave velocity v_s is known, the small strain shear modulus (G_0) can be obtained from the following expression:

$$G_0 = \rho v_s^2 \qquad (4.61)$$

4.13 SPT Correlations with Foundation Bearing Capacity

SPT results are most useful in foundation design. There are correlations between SPT N value and the base resistance for shallow foundations and, many for shaft and base resistances of foundations.

Poulos (2014) reports the work of Decourt (1995) and Table 4.11 presents correlation factors for shallow and deep foundations which could be used with the equations given below.

$$\text{Shallow foundations : ultimate bearing capacity } q_u = K_1 . N_r \text{kPa} \qquad (4.62)$$

$$\text{Piles : ultimate base resistance } f_b = K_2 N_b \text{kPa} \qquad (4.63)$$

$$\text{Piles : ultimate shaft resistance } f_s = \alpha . [2.8 N_s + 10] \text{kPa} \qquad (4.64)$$

where

N_r = average SPT (N_{60}) value within depth of one-half of the footing width
N_s = SPT value along pile shaft

Table 4.10 Some Correlations between v_s (m/s) and N

Source		Correlation
Imai (1977)	All soils	$v_s = 91N^{0.337}$
	Sand	$v_s = 80.6N^{0.331}$
	Clay	$v_s = 80.2N^{0.292}$
Ohta and Goto (1978)	All soils	$v_s = 85.35N^{0.348}$
Seed and Idriss (1981)	All soils	$v_s = 61.4N^{0.5}$
Sykora and Stokoe (1983)	Sand	$v_s = 100.5N^{0.29}$
Okamoto et al. (1989)	Sand	$v_s = 125N^{0.3}$
Pitilakis et al. (1999)	Sand	$v_s = 145N^{0.178}$
	Clay	$v_s = 132N^{0.271}$
Kiku et al. (2001)	All soils	$v_s = 68.3N^{0.292}$
Jafari et al. (2002)	Sand	$v_s = 22N^{0.77}$
	Clay	$v_s = 27N^{0.73}$
Hasancebi and Ulusay (2007)	All soils	$v_s = 99N^{0.309}$
	Sand	$v_s = 90.82N^{0.319}$
	Clay	$v_s = 97.89N^{0.269}$
Dikmen (2009)	All soils	$v_s = 58N^{0.39}$
	Sand	$v_s = 73N^{0.33}$
	Silt	$v_s = 60N^{0.36}$
	Clay	$v_s = 44N^{0.48}$

$N_b =$ SPT value close to pile tip
$K_1, K_2 =$ factors shown in Table 4.11
$\alpha = 1$ for displacement piles in all soils and non-displacement piles in clay
$\alpha = 0.5 - 0.6$ for non-displacement piles in granular soils

For pile foundations, Poulos (1989) divides analysis and design procedures for axial capacity into three categories:

Category 1: Correlations with SPT or CPT and total stress method (e.g. Tomlinson 1957)
Category 2: Effective stress method (e.g. Burland 1973; Fleming et al. 1985)
Category 3: Plasticity solutions for end bearing capacity (e.g. Meyerhof 1963), analytical and numerical solutions

Poulos (1989) states that *"Category 1 procedures probably account for most pile designs done throughout the world"*. He compiled correlations as presented in Tables 4.12 and 4.13.

4.13 SPT Correlations with Foundation Bearing Capacity

Table 4.11 Correlation factors K_1 and K_2

Soil type	K_1 Shallow footings	K_2 Displacement piles	K_3 Non-displacement piles
Sand	90	325	165
Sandy silt	80	205	115
Clayey silt	80	165	100
Clay	65	100	80

After Decourt (1995)

Table 4.12 Correlations between shaft resistance f_s and SPT N value with $f_s = \alpha + \beta N$ kN/m^2

Pile type	Soil type	α	β	Remarks	References
Driven displ.	Cohesionless	0	2.0	f_s = average value over shaft N = average SPT along shaft Halve f_s for small displacement piles	Meyerhof (1956) Shioi and Fukui (1982)
	Cohesionless & cohesive	10	3.3	Pile type not specified $50 \geq N \geq 3$ $f_s \not> 170$ kN/m^3	Decourt (1982)
	Cohesive	0	10		Shioi and Fukui (1982)
Cast in place	Cohesionless	30	2.0	$f_s \not> 200$ kN/m^3	Yamashita et al. (1987)
		0	5.0		Shioi and Fukui (1982)
	Cohesive	0	5.0	$f_s \not> 150$ kN/m^3	Yamashita et al. (1987)
		0	10.0		Shioi and Fukui (1982)
Bored	Cohesionless	0	1.0		Findlay (1984) Shioi and Fukui (1982)
		0	3.3		Wright and Reese (1979)
	Cohesive	0	5.0		Shioi and Fukui (1982)
	Cohesive	10	3.3	Piles cast under bentonite $50 \geq N \geq 3$ $f_s \not> 170$ kN/m^3	Decourt (1982)
	Chalk	−125	12.5	$30 > N > 15$ $f_s \not> 250$ kN/m^3	After Fletcher and Mizon (1984)

After Poulos (1989)

Table 4.13 Correlations between end bearing resistance f_b and SPT N value with $f_b = K N$ MN/m^2

Pile type	Soil type	K	Remarks	References
Driven displ.	Sand	0.45	N = average value in local failure zone	Martin et al. (1987)
	Sand	0.40		Decourt (1982)
	Silt, sandy silt	0.35		Martin et al. (1987)
	Glacial coarse to fine silt deposits	0.25		Thorburn and Mac Vicar (1971)
	Residual sandy silts	0.25		Decourt (1982)
	Residual clayey silts	0.20		Decourt (1982)
	Clay	0.20		Martin et al. (1987)
	Clay	0.12		Decourt (1982)
	All soils	0.30	For $L/d \geq 5$ If $L/d < 5$, $K = 0.1 + 0.04\ L/d$ (close end piles) or $K = 0.06\ L/d$ (open-ended piles)	Shioi and Fukui (1982)
Cast in place	Cohesionless		$f_b = 3.0$ MN/m^2	Shioi and Fukui (1982)
		0.15	$f_b \not> 7.5$ MN/m^2	Yamashita et al. (1987)
	Cohesive	–	$f_b = 0.09\ (1 + 0.16z)$ where z = tip depth (m)	Yamashita et al. (1987)
Bored	Sand	0.1		Shioi and Fukui (1982)
	Clay	0.15		Shioi and Fukui (1982)
	Chalk	0.25	$N < 30$	Hobbs (1977)
		0.20	$N > 40$	

After Poulos (1989)

References

American Society for Testing and Materials (2014) Annual book of ASTM standards, 04.08, West Conshohocken

Bazaraa A (1967) Use of the standard penetration test for estimating settlements of shallow foundations on sand, Ph.D. dissertation, Civil Engineering Department, University of Illinois, Champaign-Urbana

Begemann HKS (1974) General report for central and western Europe. In: Proceedings, European symposium on penetration testing, Stockholm

Bowles JE (1996) Foundation analysis and design, 5th edn. McGraw-Hill, New York

Burland JB (1973) Soft friction piles in clay – a simple fundamental approach. Ground Eng 6(3):30–42

Cubrinovski M, Ishihara K (1999) Empirical correlations between SPT N-values and relative density for sandy soils. Soils Found 39(5):61–92

References

Cubrinovski M, Ishihara K (2002) Maximum and minimum void ratio characteristics of sands. Soils Found 42(6):65–78

Das BM, Ramana GV (2011) Principles of soil dynamics, 2nd edn. Cengage, Stamford

Decourt L (1982) Prediction of the bearing capacity of piles based exclusively on N values of the SPT. In: Proceedings ESOPT II, Amsterdam, 1, pp 29–34

Decourt L (1995) Prediction of load-settlement relationships for foundations on the basis of the SPT-T. In: Ciclo de Conferencias Intern. "Leonardo Zeevaert", UNAM, Mexico, pp 85–104

DeMello VFB (1971) Standard penetration test. In: Proceedings, 4th Panamerican conference on soil mechanics and foundation engineering, San Juan, Puerto Rico, ASCE, pp 1–86

Dikmen Ü (2009) Statistical correlations of shear wave velocity and penetration resistance for soils. J Geophys Eng 6(1):61–72

Ferrent TA (1963) The prediction of field verification of settlement on cohesionless soils. In: Proceedings, 4th Australia-New Zealand conference on soil mechanics and foundation engineering, pp 11–17

Findlay JD (1984) Discussion. In: Piling and ground treatment. Institution of Civil Engineers, Thomas Telford, London, pp 189–190

Fleming WGK, Weltman AJ, Randolph MF, Elson WK (1985) Piling Engineering. Surrey University Press/Halsted Press, New York

Fletcher MS, Mizon DH (1984) Piles in chalk for Orwell bridge. In: Piling and ground treatment. Institution of Civil Engineers, Thomas Telford, London, pp 203–209

Gibbs HJ, Holtz WG (1957) Research on determining the density of sand by spoon penetration testing. In: Proceedings, 4th international conference on soil mechanics and foundation engineering, London, I, pp 35–39

Hasancebi N, Ulusay R (2007) Empirical correlations between shear wave velocity and penetration resistance for ground shaking assessments. Bull Eng Geol Environ 66(2):203–213

Hatanaka M, Uchida A (1996) Empirical correlations between penetration resistance and internal friction angle of sandy soils. Soils Found 36(4):1–10

Hobbs NB (1977) Behavior and design of piles in chalk – an introduction to the discussion of the papers on chalk. In: Proceedings, symposium on piles on weak rock, London, pp 149–175

Holtz WG, Gibbs HJ (1979) Discussion of "SPT and relative density in coarse sand". J Geotech Eng Div ASCE 105(3):439–441

Imai T (1977) P- and S-wave velocities of the ground in Japan. In: Proceedings, 9th international conference on soil mechanics and foundation engineering, Tokyo, Japan, 2, pp 257–260

Jafari MK, Shafiee A, Razmkhah A (2002) Dynamic properties of the fine grained soils in south of Tehran. J Seismol Earthquake Eng 4(1):25–35

Kiku H, Yoshida N, Yasuda S, Irisawa T, Nakazawa H, Shimizu Y, Ansal A, Erkan A (2001) In-situ penetration tests and soil profiling in Adapazari, Turkey. In: Proceedings, 15th international conference on soil mechanics and geotechnical engineering, TC4 Satellite Conference on Lessons Learned from Recent Strong Earthquakes, Istanbul, Turkey, pp 259–269

Kishida H (1966) Damage to reinforced concrete buildings in Niigata City with special reference to foundation engineering. Soils Found 7(1):75–92

Kuizumi Y (1966) Changes in density of sand subsoil caused by the Niigata earthquake. Soils Found 8(2):38–44

Kulhawy FH, Mayne PW (1990) Manual on estimating soil properties for foundation design. Electric Power Research Institute, Palo Alto

Liao SSC, Whitman RV (1986) Overburden correction factors for SPT in sand. J Geotech Eng ASCE 112(3):373–377

Marcuson WF III, Bieganousky WA (1977) SPT and relative density in coarse sands. J Geotech Eng Div ASCE 103(11):1295–1309

Martin RE, Seli JJ, Powell GW, Bertoulin M (1987) Concrete pile design in Tidewater, Virginia. J Geotech Eng ASCE 113(6):568–585

Mayne PW, Kemper JB (1988) Profiling OCR in stiff clays by CPT and SPT. Geotech Test J ASTM 11(2):139–147

Meyerhof GG (1956) Penetration tests and bearing capacity of cohesionless soils. J Soil Mech Found Eng ASCE 82(SM1):1–19

Meyehof GG (1957) Discussion on research on determining the density of sand by spoon penetration testing. In: Proceedings, 4th international conference on soil mechanics and foundation engineering, London, 3, pp 110–114

Meyerhof GG (1959) Compaction of sands and the bearing capacity of piles. J Soil Mech Found Div ASCE 85(6):1–29

Meyerhof GG (1963) Some recent research on the bearing capacity of foundations. Can Geotech J 1:16–26

Mitchell JK, Gardner WS (1975) In situ measurement of volume characteristics. In: Proceedings, specialty conference, geotechnical engineering division, ASCE 2, pp 279–345

Ohasaki Y (1966) Niigate earthquake 1964, building damage and soil conditions. Soils Found 6(2):14–37

Ohta Y, Goto N (1978) Empirical shear wave velocity equations in terms of characteristic soil Iindexes. Earthquake Eng Struct Dyn 6:167–187

Okamoto T, Kokusho T, Yoshida Y, Kusuonoki K (1989) Comparison of surface versus subsurface wave source for P-S logging in sand layer. In: Proceedings, 44th annual conference, Japan society of civil engineers, 3, pp 996–997 (in Japanese)

Peck RB, Hanson WE, Thornburn TH (1974) Foundation engineering, 2nd edn. Wiley, New York

Pitilakis K, Raptakis D, Lontzetidis KT, Vassilikou T, Jongmans D (1999) Geotechnical and geophysical description of euro-seistests using field and laboratory tests, and moderate strong ground motions. J Earthquake Eng 3:381–409

Poulos HG (1989) Pile behaviour–theory and application. Geotechnique 39(3):365–415

Poulos HG (2014) Tall building foundations – design methods and applications. In: Proceedings conference XXVII RNIG, Sociedad Mexicana de Ingeniería Geotécnia A.C., Puerto Vallarta, Jalisco

Robertson PK, Wride CE (1998) Evaluating cyclic liquefaction potential using the cone penetrometer test. Can Geotech J 35(3):442–459

Salgado R (2008) The engineering of foundations. McGraw-Hill, New York

Salgado R, Boulanger R, Mitchell JK (1997) Lateral stress effects on CPT liquefaction resistance correlations. J Geotechn Geoenviron Eng 123(8):726–735

Schmertmann JH (1975) Measurement of *in situ* shear strength. In: Proceedings, specialty conference on in situ measurement of soil properties, ASCE, 2, pp 57–138

Seed HB (1979) Soil liquefaction and cyclic mobility evaluation for level ground during earthquakes. J Geotech Eng Div ASCE 105(2):201–255

Seed HB, Idriss IM (1981) Evaluation of liquefaction potential of sand deposits based on observations and performance in previous earthquakes, Preprint No. 81–544, In situ testing to evaluate liquefaction susceptibility, Session No. 24, ASCE Annual Conference, St. Louis, Missouri

Seed HB, Arango I, Chan CK (1975) Evaluation of soil liquefaction potential during earthquakes, Report No. EERC 75–28. Earthquake Engineering Research Center, University of California, Berkeley

Seed HB, Tokimatsu K, Harder LF, Chung RM (1985) Influence of SPT procedures in soil liquefaction resistance evaluations. J Geotech Eng ASCE 111(12):1425–1445

Shioi Y, Fukui J (1982) Application of N-value to design of foundations in Japan. In: Proceedings 2nd European symposium on penetration testing, Amsterdam, 1, pp 159–164

Skempton AW (1986) Standard penetration test procedures and the effect in sands of overburden pressure, relative density, particle size, aging and overconsolidation. Geotechnique 36(3): 425–447

Stroud MA (1975) The standard penetration test in insensitive clays and soft rocks. In: Proceedings, European symposium on penetration testing 2, pp 367–375

Sykora DE, Stokoe KH (1983) Correlations of in-situ measurements in sands of shear wave velocity. Soil Dyn Earthquake Eng 20:125–136

Szechy K, Varga L (1978) Foundation engineering – soil exploration and spread foundation. Akademiai Kiado, Hungary
Terzaghi K, Peck RB (1967) Soil mechanics in engineering practice, 2nd edn. Wiley, New York
Tokimatsu K, Yoshimi Y (1983) Empirical correlation of soil liquefaction based on SPT N-value and fines content. Soils Found 23(4):56–74
Tomlinson MJ (1957) The adhesion of piles driven into clay soils. In: Proceedings 4th international conference soil mechanics and foundation engineering, London, 2, pp 66–71
Thornburn S, MacVicar RS (1971) Pile load tests to failure in the Clyde alluvium. In: Proceedings, conference on behaviour of piles, London, pp 1–7
Webb DL (1969) Settlement of structures on deep alluvial sandy sediments in Durban, South Aftica. In: Proceedings, conference on the in situ behavior of soils and rocks, Institution of Civil Engineering, London, pp 181–188
Wolff TF (1989) Pile capacity prediction using parameter functions. In: Predicted and observed axial behavior of piles, results of a pile prediction symposium, ASCE Geotechnical Special Publication 23, pp 96–106
Wright SJ, Reese LC (1979) Design of large diameter bored piles. Ground Eng 12(8):47–51
Yamashita K, Tomono M, Kakurai M (1987) A method for estimating immediate settlement of piles and pile groups. Soils Found 27(l):61–76
Yoshida Y, Ikemi M, Kokusho T (1988) Empirical formulas of SPT blow-counts for gravelly soils. In: Proceedings, 1st international symposium on penetration testing, Orlando, Florida, 1, pp 381–387

Chapter 5
Cone Penetrometer Test

Abstract This chapter covers one of the most popular insitu tests, the cone penetrometer test (CPT). A brief description of different types of tests has been provided. The piezocone test, an advanced CPT, is described including the test procedure and the parameters obtained in the field. Pore pressure transducer placement locations and the corrections to measured pore pressures are presented. A detailed discussion of soil classification using cone test results is provided. Correlations for design parameters related to sands and clays are discussed separately. For sands, correlations include relative density, friction angle, modulus and small strain modulus. For clays, correlations include the undrained shear strength, over consolidation ratio, constrained modulus, small strain shear modulus, compressibility, friction angle, unit weight and permeability. As the CPT test competes with SPT test for popularity, correlations between the two tests are also discussed. Correlations to use CPT derived parameters directly to calculate the ultimate bearing capacity of shallw and deep foundations are presented. The chapter concludes with a section on liquefaction assessment using CPT as well as SPT results.

Keywords Cone penetrometer test · CPT · Correlations · Piezocone · Soil classification · Liquefaction · Bearing capacity

5.1 Cone Penetrometer Test – General

The cone penetrometer test (CPT) is a versatile in situ test (Fig. 5.1) which has become a routine test for site investigations worldwide to characterize clays and sands. There is little doubt that the cone penetrometer test is one of the the most widely used in situ test in areas where soft and compressible soils occur. As the test is a continuous test, the subsoil profile variation is captured with significantly more details compared to a vane shear test or a SPT which are generally carried out at 1–1.5 m depth intervals. It is a test most useful in weak clays and sands. Latest machinery used for advancing CPT in soils have more power and robustness compared to early equipment and therefore its use in competent soils such as very stiff to hard clays and dense sands is generally not an issue. Modern advances on CPT rigs allow the recovery of undisturbed samples or carry out vane shear tests in addition to carrying out conventional CPT testing. This is very advantageous

Fig. 5.1 CPT machine in mud flats (Courtesy Yvo Keulemans, CPTS)

because of significant additional costs involved if a separate borehole rig has to be mobilised for sampling and vane shear testing.

The main disadvantage of the CPT is that it does not provide an absolute value for soil parameters and the results need to be calibrated against other tests such as vane shear and laboratory tests such as triaxial tests. Where such data are not available, practitioners use local experience and/or empirical values to derive design parameters

There are generally three main types of cone penetrometers:

1. Mechanical cone penetrometer – Also known as the Dutch Cone Penetrometer or the Static Cone Penetrometer, this uses a set of solid rods or thick walled tubes to operate the penetrometer. The penetrometer tip is initially pushed about 4 cm and the tip resistance is recorded. Then both cone and sleeve are pushed together to record the combined tip and cone resistance. This is repeated with depth to provide a profile for cone and sleeve resistances. The procedure allows a measurement be taken at about every 20 cm.
2. Electric cone penetrometer – An advancement of the mechanical cone, the electric cone has transducers to record the tip and sleeve resistances separately. Therefore it has the advantage of advancing the cone continuously to obtain a continuous resistance profile and the inner rods are not required.
3. Electric cone penetrometer with pore pressure measurements (Piezocone) – A further addition to the penetrometer is the inclusion of pore pressure transducers at the tip or on the sleeve to record continuous pore pressure measurements. The test is widely known with the abbreviation CPTu. A more popular name for the CPTu equipment is the piezocone. In Chap. 1, Table 1.1, where applicability and usefulness of in situ tests are summarised, it is evident that the piezocone has the best rating amongst in situ tests for parameters obtained or the ground type investigated.

A further development of penetrometer type is the seismic cone penetrometer test which allows the measurement of the shear wave velocity with depth. The equipment consists of the piezocone unit plus a receiver for seismic measurements. Generally, at every 1 m interval (i.e. at rod breaks), a shear wave is generated at the ground surface and the seismic wave arrival time is recorded. The shear wave velocity could be converted to a shear modulus (see Sect. 5.5.4) using empirical correlations. The seismic cone penetrometer test is not addressed here and the reader is referred to Mayne (2007).

The main difference between the traditional CPT and the CPTu is the measurement of pore pressure in the latter test. The measurement of pore pressure provides a wealth of data, with a pore pressure profile at the test location, reflecting the different soil types. This type of data was never available to the designer as no other test in history could measure continuous pore pressure with depth. CPTu therefore became very popular within a relatively short time although traditional CPTs are still in use, probably because of the cost of new equipment and accessories. Most likely, the traditional CPT will be phased out from the market, especially in the developed world, because of the advantages offered by the CPTu.

Equations and empirical relationships available for the CPT are equally valid for the CPTu with additional relationships established due to extra information provided by pore pressure measurements. Therefore, in this chapter, we will be referring to the CPTu rather than the CPT although correlations that do not include pore pressure measurements would still be valid for both.

One of the other achievements in the CPTu is its ability to carry out dissipation tests to obtain the coefficient of consolidation of clays. At a nominated depth, advancement of the cone is stopped and the pore pressure generated is allowed to dissipate and the measurements are recorded continuously. Generally a target of 50 % dissipation is adopted because of time constraints. Even such a limited duration in soft soils could be in the order of an hour or two. Where the dissipation is very slow because the coefficient of consolidation, c_v, is very low, some operators leave the test overnight for dissipation. Readers are referred to Lunne et al. (1997) for a description of the test and derivation of geotechnical parameters related to the rate of consolidation.

5.2 Piezocone Test – Equipment and Procedure

ISSMGE Technical Committee 16 (TC16) (Ground Characterization from In Situ Testing) published an International Test Procedure for the CPT and the CPTu. The information given below on the equipment (see Fig. 5.2) and procedure is mostly based on that report.

The piezocone test consists of a cone and a surface sleeve continuously pushed into the ground and the resistance offered by the cone and sleeve measured

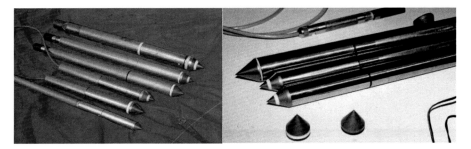

Fig. 5.2 Various Cone Penetrometers including Electric Friction and Piezocone types (After FHWA NHI-01-031 – Mayne et al. 2002)

electronically, in addition to measuring the pore pressure by the use of a pore pressure transducer. The standard cone tip is usually 10–15 cm^2 and has an apex angle of 60°). The cone is pushed with a standard rate of penetration of 20 ± 5 mm/s. Figure 5.3 shows the location of main components i.e. cone, sleeve and the pore pressure transducer of the probe which is pushed down by rods. The measurements taken with depth include:

- Tip resistance (q_c)
- Sleeve friction (f_s)
- Pore water pressure (u) – could be measured at the cone face (u_1), shoulder (u_2) or top of the sleeve (u_3) (see Fig. 5.3)

Tip resistance (q_c) is obtained by measuring the ultimate force (Q_c) experienced by the cone only divided by the area of the cone (A_c).

$$q_c = \frac{Q_c}{A_c} \quad (5.1)$$

Sleeve friction (f_s) is obtained by measuring the ultimate force (Q_s) only on the sleeve, i.e., side friction, divided by the area of the sleeve (A_s).

$$f_s = \frac{Q_s}{A_s} \quad (5.2)$$

In piezocones/CPTu's, the pore pressure transducer is located generally at mid-face of the cone (measuring u_1) or at the shoulder (measuring u_2) i.e. where cone and sleeve meet (see Fig. 5.2). The resistance measurements are influenced by the water pressure acting behind the cone tip and the edge of the sleeve. Therefore a correction needs to be applied.

When the pore pressure transducer is located at the shoulder, the following equation could be used to correct the cone resistance (Jamiolkowski et al. 1985; Campanella and Robertson 1988; Lunne et al. 1997; Campanella et al. 1982; Mayne 2007):

Fig. 5.3 Components and correction details of a piezocone (Adapted from FHWA NH1-01-031 – Mayne et al. 2002)

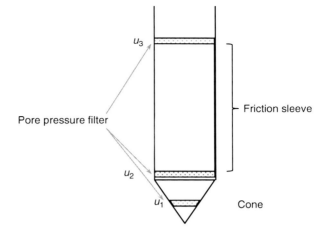

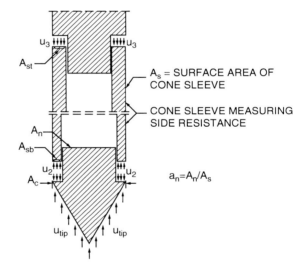

$$q_t = q_c + u_2(1 - a_n) \tag{5.3}$$

where

q_t = corrected tip resistance
u_2 = pore pressure measured at the shoulder
a_n = net area ratio, approximately equal to the ratio of shaft cross section and cone cross section areas (usually measured in a calibration cell, Lunne et al. 1997)

As Lunne et al. (1997) points out, although a value close to 1.0 is ideal, the ratio a_n generally ranges from 0.55–0.9. However, values as low as 0.38 have been recorded which should be unacceptable when the test is carried out in very soft fine grained soils because the correction becomes the main contribution to q_t.

A correction factor is also applicable to the sleeve friction (Lunne et al. 1997):

$$f_t = f_s + \frac{(u_3 A_{st} - u_2 A_{sb})}{A_s} \qquad (5.4)$$

where

A_{sb} = cross sectional area of the sleeve at the base – Fig. 5.3.
A_{st} = cross sectional area of the sleeve at the top – Fig. 5.3.
A_s = surface area of the sleeve – Fig. 5.3.

In clayey soils, the magnitude of pore water pressures generated during a test could be high and the correction can be significant. Therefore, the correction factor is most important to correct the recorded tip resistance and provide more accurate results. However, pore water pressure correction is not important for sands because the pore pressure generated is not significant and therefore the measured pore pressure purely reflects the height of the groundwater table i.e, measures the hydrostatic pressure. Therefore it is not significant whether q_t or q_c is used for engineering assessments when sandy soils are present.

Although the ISSMGE TC16 (1999) Reference Test Procedure refers to pore pressure measurement at the shoulder, i.e. u_2, some penetrometers measure the pore pressure at the cone (u_1) in which case u_2 could be obtained as follows (Lunne et al. 1997):

$$(u_2 - u_0) = K(u_1 - u_0) \qquad (5.5)$$

where u_0 = equilibrium pore pressure (due to groundwater table)

Typical values for K presented by Lunne et al. 1997 (modified after Sandevan 1990) are presented in Table 5.1.

Two other important parameters related to CPTu and will be discussed later are the Friction Ratio (R_f) and the pore water pressure parameter (B_q). R_f is deduced from the two parameters q_t and f_t:

$$R_f = \frac{f_t}{q_t} \qquad (5.6)$$

B_q is derived from the following equation:

$$B_q = \frac{u_2 - u_o}{q_t - \sigma_{vo}} \qquad (5.7)$$

where

σ_{vo} = total overburden stress

Table 5.1 Typical values for adjustment factor K if filter is located at the cone (Lunne et al. 1997)

Soil type	K	u_1/u_0
CLAY normally consolidated	0.6–0.8	2–3
CLAY slightly overconsolidated, sensitive	0.5–0.7	6–9
CLAY heavily overconsolidated, stiff	0–0.3	10–12
SILT loose, compressible	0.5–0.6	3–5
SILT dense, dilative	0–0.2	3–5
SAND loose, silty	0.2–0.4	2–3

5.3 Practical Use of Penetrometer Test Results

The cone penetrometer test is a complex test to be analysed and it is largely used with empirical relationships for all practical purposes. In addition to soil classification and provision of a continuous profile the test allows the derivation of several geotechnical parameters. The main uses of the test for the practitioners could be summarized as follows:

1. Soil classification
2. Correlations for Cohesionless soils

    (a) Relative density
    (b) Friction angle
    (c) Modulus
    (d) Small strain shear modulus

3. Correlations for Cohesive soils

    (a) Undrained Shear Strength
    (b) Sensitivity
    (c) Over consolidation ratio (OCR)
    (d) Modulus and compressibility
    (e) Small strain shear modulus
    (f) Friction angle

4. Correlation with unit weight
5. Correlation with foundation resistance
6. Correlation with SPT
7. Correlation with permeability

5.4 Soil Classification

One of the primary objectives of a cone penetrometer test is to identify the soil profile from the test results. While empirical rules have been established for this purpose as discussed later, it should be stressed that a probe test can never displace/replace borehole sampling which allows physical observation of the materials and allows laboratory tests to be carried out. Therefore it is generally considered good practice to conduct boreholes to supplement the CPT programme so that CPT results could be calibrated against and also to collect samples for laboratory testing.

Initially, calibration of penetrometer test results were carried out using the cone resistance and friction ratio as pore pressure measurements were not available until the advent of the piezocone. One of the earliest comprehensive classifications was carried out by Douglas and Olsen (1981) using cone resistance and the friction ratio. It is generally accepted that the measurement of sleeve friction is often less accurate and less reliable than the cone resistance.

The advent of the piezocone allowed the additional measurement of pore pressure which allows better classification of soils. The main additional parameter used for soil classification is the pore pressure ratio B_q [Eq. (5.7)].

Over the years, many an author has proposed classification methods and charts based on either CPT or CPTu test results. They include Schmertmann (1978), Robertson et al. (1986), Robertson (1990), Eslami and Fellenius (1997), Olsen and Mitchell (1995), Senneset et al. (1989), Jones and Rust (1982), Ramsey (2002) and Jefferies and Davies (1991). Long (2008) reviewed the proceedings of various conferences to find out the commonly used classification charts by academics, researchers and practitioners. He concluded that, over the period 1998–2006, Robertson et al. (1986) and Robertson (1990) charts are the most popular. Long (2008) also reports on Molle (2005) who carried out a review of published literature to determine the reliability of different classification charts and concluded that Robertson et al. (1986) and Robertson (1990) charts provided reasonable to very good results. Therefore only these two methods are briefly discussed below.

The two methods of Robertson et al. (1986) and Robertson (1990) use the following parameters for the classification charts:

- Robertson et al. (1986) – q_t, B_q and R_f (Fig. 5.4)
- Robertson (1990) – Q_t (normalized q_t), B_q and F_r (normalized friction ratio) (Fig. 5.5)

where

$$\text{Normalized cone resistance } Q_t = \frac{q_t - \sigma_{vo}}{\sigma'_{vo}} \qquad (5.8)$$

$$\text{Normalized friction ratio } F_r = \frac{f_s}{q_t - \sigma_{v0}} 100\,(\%) \qquad (5.9)$$

$$\text{effective vertical stress } \sigma'_{vo} = \sigma_{vo} - u_0 \qquad (5.10)$$

From the two charts in Fig. 5.5, only the first chart could be used if a CPT is carried out without pore pressure measurements.

Robertson et al. (1986) state that linear normalization of cone resistance (Q_t) is best suited for clay soils and less appropriate for sands.

5.4 Soil Classification

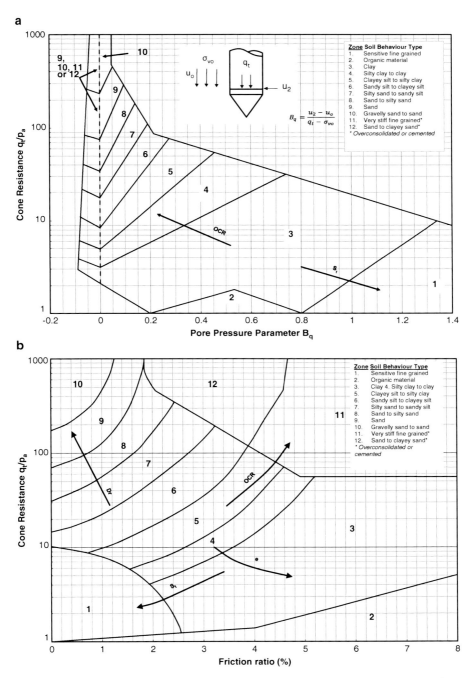

Fig. 5.4 Proposed soil behaviour type classification system from CPTu data (Adapted from Robertson et al. 1986)

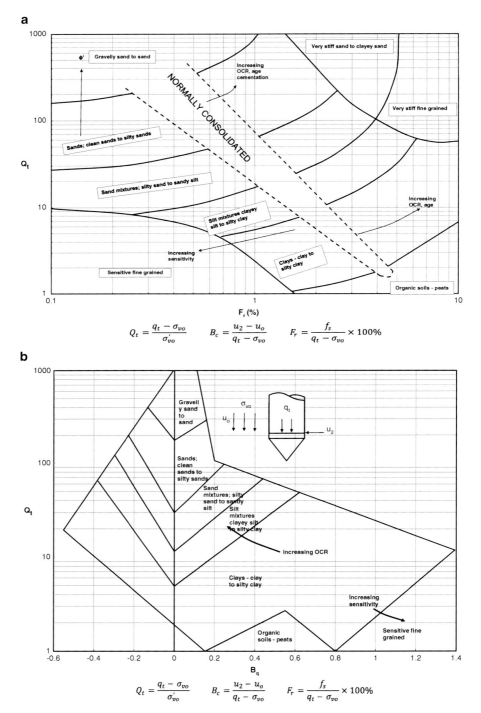

Fig. 5.5 Soil behaviour type classification chart based on normalised CPT/CPTu data (adapted from Robertson 1990)

5.5 Correlations for Sands

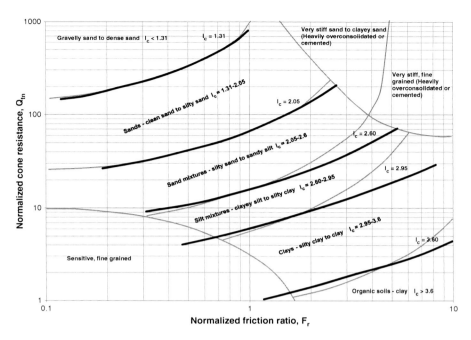

Fig. 5.6 Contours of SBT index, I_c on normalized SBT charts of Robertson (1990)
Note – Normally Consolidated Zone (see Fig. 5.5) left out for clarity but should be superimposed

Jefferies and Davies (1993) introduced an index I_c to represent Soil Behaviour Type (SBT) zones in the Q_t vs F_r chart (Fig. 5.5) of Robertson (1990). The index I_c is the radius of circle defining the zone boundaries. Robertson and Wride (1998) presented a modification to this index that could be applied to the chart in Fig. 5.5. The modified definition is:

$$I_c = \left[(3.47 - \log Q_t)^2 + (\log F_r + 1.22)^2\right]^{0.5} \quad (5.11)$$

Figure 5.6 shows the modified chart of Fig. 5.5.

5.5 Correlations for Sands

5.5.1 *Correlation with Relative Density of Sand*

Relative density, D_r, is a parameter used in sands to identify the level of compaction of the material (see Sect. 2.3.2). It is given by the formula:

$$D_r = \frac{e_{max} - e}{e_{max} - e_{min}} \qquad (5.12)$$

where

e_{max} = maximum void ratio
e_{min} = minimum void ratio
e = in situ void ratio

This equation can be re-written as follows:

$$D_r = \frac{(\rho_d - \rho_{d\,min})}{(\rho_{d\,max} - \gamma_{d\,min})} \frac{\rho_{d\,max}}{\rho_d} \qquad (5.13)$$

where

$\rho_{d\,max}$ maximum dry density
$\rho_{d\,min}$ minimum dry density
ρ_d in situ dry density

Relative Density could be evaluated using laboratory and field tests specifically catered to obtain density values:

1. Use laboratory procedures to find minimum and maximum density, and e_{max} and e_{min}
2. Use in situ tests such as sand replacement method and nuclear gauge test to find the in situ dry density and hence e.

There are various country standards detailing the laboratory procedures to obtain the maximum and minimum density including ASTM Standards D4253-14 and D4254-14. However, the in situ density using sand replacement or nuclear gauge can only be measured at shallow depth because it is not practical or feasible to excavate test pits more than about 1 m deep to carry out such tests. The use of CPT and SPT type tests are used widely in the industry to assess relative density by the use of empirical rules to convert the penetration resistance.

Early research has categorized the density to describe the relative behaviour as shown in Table 5.2.

Baldi et al. (1986) presented several correlations for both normally consolidated (NC) and over consolidated (OC) sands. The following correlation is for NC sand:

$$D_r = \left(\frac{1}{C_2}\right) \ln\left(\frac{q_c}{C_0 (\sigma'_{vo})^{0.55}}\right) \qquad (5.14)$$

where:

C_0 and C_2 = soil constants (For moderately compressible, normally consolidated, unaged and uncemented, predominantly quartz sands, $C_0 = 157$ and $C_2 = 2.41$)

5.5 Correlations for Sands

Table 5.2 Borderline values of D_r, N and ϕ' for granular soils

	Very loose		Loose		Medium dense		Dense		Very dense
#D_r (%)	0		15		35		65		85	100
˙N_{60}			4		10		30		50	
##$(N_1)_{60}$			3		8		25		42	
˙˙ϕ'(deg)			28		30		36		41	
##$(N_1)_{60}/D_r^2$					65		59		58	

˙Terzaghi and Peck (1948); #Gibbs and Holtz (1957); ##Skempton (1986); ˙˙Peck et al. (1974)

q_c = cone penetration resistance in kPa
σ'_{vo} = effective vertical stress in kPa

Mayne (2007) states that most correlations in the 70's and 80's did not consider the boundary effects of the calibration chambers and refers to Jamiolkowski et al. (2001) who introduced a correction factor in reexamining previous results and the expression proposed was as follows:

$$D_r = 100 \left[0.268 \ln \left(\frac{q_t/p_a}{\sqrt{\sigma'_{vo}/p_a}} \right) - 0.675 \right] \quad (5.15)$$

Another widely used relationship was proposed by Kulhawy and Mayne (1990) who highlighted the effects of compressibility and over consolidation ratio (OCR) on the relationship between the relative density and the dimensionless cone resistance. Based on available corrected calibration studies they proposed the following approximate solution to capture the different relationships:

$$D_r^2 = \frac{Q_{cn}}{305 \, Q_c \, Q_{OCR}} \quad (5.16)$$

where

$Q_{cn} = (q_c/p_a) / (\sigma'_v/p_a)^{0.5}$
Q_C = Compressibility factor (0.91 for high, 1.0 for medium and 1.09 for low). Kulhawy and Mayne (1990) state that majority of the natural sands are likely to be of medium to high compressibility
Q_{OCR} = Overconsolidation factor = $OCR^{0.18}$
p_a = Atmospheric pressure in same units as q_c
σ'_v = effective vertical stress in same units as q_c

Table 5.3 Values of coefficients a and c in Eq. (5.19)

Reference	a	c	Remarks
Burland and Burbidge (1985)	0.33	15.49	Upper limit
	0.32	4.90	Lower limit
Robertson and Campanella (1983)	0.26	10	Upper limit
	0.31	5.75	Lower limit
Kulhawy and Mayne (1990)	0.25	5.44	
Canadian Geotechnical Society (1992)	0.33	8.49	
Anagnostopoulos et al. (2003)	0.26	7.64	

The above equation is similar in form to the equation Kulhawy and Mayne (1990) proposed for the SPT with the most important difference being SPT relationship included aging whereas the above CPT relationship is only for unaged sands. They suggested that, if the same functional relationship for aging holds for both the SPT and CPT, then it is necessary to include the aging factor they proposed for the SPT relationship. This results in the following equation although they warn that this addition is currently speculative:

$$D_r^2 = \frac{Q_{cn}}{305\, Q_c\, Q_{OCR}\, Q_A} \quad (5.17)$$

where

$$Q_A = \text{Aging factor} = 1.2 + 0.05 \log(t/100),\ t \text{ in years} \quad (5.18)$$

Das and Sivakugan (2011) summarise the work of several authors who investigated the relationship between q_c, N_{60} (SPT at 60 % efficiency) and D_{50} (median grain size). The correlations can be expressed as follows:

$$\frac{\left(\frac{q_c}{p_a}\right)}{N_{60}} = c D_{50}^a \quad (5.19)$$

Table 5.3 lists the average values for a and c from these studies.

Further relationships between SPT N and q_c are presented in Sect. 5.9.

5.5.2 Correlation of q_c with Sand Friction Angle, ϕ'

There have been several attempts to interpret the friction angle of sand from CPT, specifically the CPT tip resistance, q_c (Janbu and Senneset 1974; Durgunoglu and Mitchell 1975; Villet and Mitchell 1981). One of the earliest contributions was by Meyerhoff (1956) who presented the following Table 5.4 based on the Static Cone Penetrometer.

5.5 Correlations for Sands

Table 5.4 Correlation of q_c and relative density with friction angle for cohesionless soils (After Meyerhoff 1956)

State of packing	Relative Density	SPT N	q_c in MPa	Approximate triaxial friction angle (degrees)
Very loose	<0.2	<4	<2	<30
loose	0.2–0.4	4–10	2–4	30–35
Medium dense	0.4–0.6	10–30	4–12	35–40
dense	0.6–0.8	30–50	12–20	40–45
very dense	>0.8	>50	>20	>45

Table 5.5 Correlation of q_c and relative density with friction angle for cohesionless or mixed soils (After Bergdahl et al. 1993)

Relative density	q_c (MPa)	φ' (degrees)
very weak	0.0–2.5	29–32
weak	2.5–5.0	32–35
medium	5.0–10.0	35–37
large	10.0–20.0	37–40
very large	>20.0	40–42

Robertson and Campanella (1983) proposed an empirical relationship to be applicable to uncemented, unaged, moderately compressible quartz sands after reviewing calibration chamber test results and comparing with peak friction angle from drained triaxial tests. The relationship was presented as a graph of $\log(q_c/\sigma'_{vo})$ against $\tan\phi'$. The design chart has been approximated to the following (Robertson and Cabal 2012):

$$\tan\phi' = \frac{1}{2.68}\left[\log\left(\frac{q_c}{\sigma'_{vo}}\right) + 0.29\right] \tag{5.20}$$

Dysli and Steiner (2011) cite the contribution by Bergdahl et al. (1993) as shown in Table 5.5 correlating the friction angle with q_c and relative density, for granular soils.

Mayne (2007) cites the following correlation based on the results obtained in calibration chambers:

$$\phi' = 17.6 + 11.0\log(q_{t1}) \tag{5.21}$$

where

$$q_{t1} = \frac{(q_{ct}/p_a)}{(\sigma'_{vo}/p_a)^{0.5}} \tag{5.22}$$

5.5.3 Correlation with Constrained Modulus of Cohesionless Soils

As previously discussed, to obtain undisturbed samples in cohesionless soils is difficult and expensive. Therefore, practitioners favour the use of in situ testing to determine the deformation properties by the use of empirical correlations. At shallow depth, tests such as the plate load test provide a simple but effective test although this test cannot be used to assess the deeper profile. Trial embankments could provide good evaluation of the modulus along the depth profile if adequate instrumentation such as settlement plates and extensometers are used. Obviously such tests are very expensive and may not be justified when only a limited budget is available for the site investigation. CPT, presuremeter test and dilatometer test therefore become an important tool available to the geotechnical designer.

As Baldi et al. (1989) state, although the stiffness of cohesionless soils depends on many factors including the grading, mineralogy, angularity, grain fabric, stress-strain history, mean effective stress, drainage conditions etc., in a given soil, penetration resistance is primarily controlled by the void ratio/relative density and the state of the effective stress. Based on a large number of tests carried out in situ and in a calibration chamber, proposed correlations to obtain the drained Young's modulus of silica sands based on cone penetration resistance is shown in Fig. 5.7 (Bellotti et al. 1989).

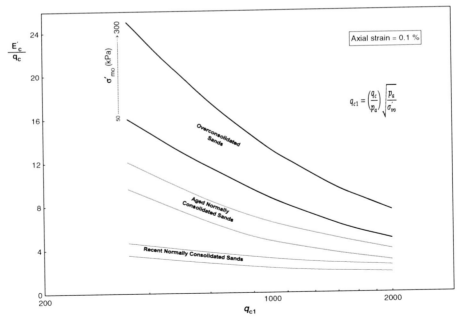

Fig. 5.7 Evaluation of drained Young's modulus from CPT for silica sands (Adapted from Bellotti et al. 1989)

5.5 Correlations for Sands

Table 5.6 Initial tangent constrained modulus correlation with q_c (Reported by Lunne et al. 1997)

	Constrained modulus (M_o) relationship with q_c (MPa)	Applicable q_c range
NC Unaged	$M_o = 4 q_c$	$q_c < 10$ MPa
	$M_o = 2 q_c + 20$ (MPa)	$10 \text{MPa} < q_c < 50$ MPa
	$M_o = 120$ MPa	$q_c > 50$ MPa
OC	$M_o = 5 q_c$	$q_c < 50$ MPa
	$M_o = 250$ MPa	$q_c > 50$ MPa

Note – M_o value represents the modulus at the insitu effective vertical stress, σ'_{vo}, before the start of in situ test

As Lunne et al. (1997) points out, most correlations between penetrometer test results and the drained constrained modulus refer to the tangent modulus as found from oedometer tests. The correlations are typically represented as:

$$M = \alpha q_c \quad (5.23)$$

where M = Constrained modulus and α = constrained modulus factor, a constant

Robertson and Campanella (1983), summarizing the work of Lunne and Kleven (1981) who reviewed calibration chamber results of different authors, commented that the results indicate an α of 3 should provide the most conservative estimate of 1-D settlement and that the choice of α value depends on judgment and experience.

Lunne et al. (1997) presented the work of Lunne and Christophersen (1983) for in situ tangent modulus (M_0) for unaged, uncemented, predominantly silica sands (see Table 5.6). As noted by Lunne et al. (1997), these correlations were based on tests carried out to a level of axial strain equal to 0.1 %, corresponding to the upper limit of the average vertical strain of practical interest of shallow and deep foundations in cohesionless soils.

To calculate modulus for higher stress ranges, Lunne and Christophersen (1983) recommended Janbu's (1963) formulation (see Lunne et al. 1997):

$$M = M_o \sqrt{\frac{\left(\sigma'_{vo} + \frac{\Delta \sigma_v}{2}\right)}{\sigma'_{vo}}} \quad (5.24)$$

where

$\Delta \sigma'_{vo}$ = additional stress above the initial stress

Robertson and Cabal (2014) suggested the following formulation:

$$M = \alpha_M (q_t - \sigma_{v0}) \quad (5.25)$$

When $I_c < 2.2$ (coarse grained soils):

$$\alpha_M = 0.0188 \left[10^{(0.55 I_c + 1.68)} \right] \qquad (5.26)$$

I_c has been defined in Eq. (5.11).

5.5.4 Correlation with Small Strain Shear Modulus of Cohesionless Soils

Small strain shear modulus, G_0, is considered valid for very small strain levels up to 0.001 %. It is generally recognized that the most appropriate way of assessing G_0 is by measuring the shear wave velocity, V_s. G_0 could then be calculated as follows:

$$G_0 = \rho V_s^2 \qquad (5.27)$$

where ρ = bulk density

However, unless project specific detailed investigations are carried out, it is generally the practice to use penetration tests to assess G_0 using empirical correlations.

Lunne et al. (1997) reports the correlation proposed by Rix and Stokes (1992) for uncemented quartz sands, shown in Eq. (5.28) and Fig. 5.8, which is based on calibration chamber test results.

$$\left(\frac{G_0}{q_c} \right)_{ave} = 1634 \left(\frac{q_c}{\sqrt{\sigma'_{vo}}} \right)^{-0.75} \qquad (5.28)$$

where G_0, q_c and σ'_{vo} are given in kPa in the range equal to Average $\pm$Average/2.

Schnaid (2009) presented a theoretical relationship for G_0 against q_c of unaged cemented soils as shown in Fig. 5.9. Together with the theoretically derived database, Schnaid (2009) has shown empirically established upper and lower bounds on the same plot.

Robertson and Cabal (2014) state that available empirical correlations to interpret in situ tests apply to unaged, uncemented silica sands. To use such equations for other sands could lead to erroneous assessments in parameters, in the current case, G_0. Therefore they proposed the following lower and upper boundaries to characterize uncemented, unaged sands based on Eslaamizaad and Robertson (1997):

$$G_0 = b \left(q_t \sigma'_{vo} P_a \right)^{0.3} \qquad (5.29)$$

where 'b' is a constant of 280 for upper bound and 110 for lower bound.

It is a narrow range and if the results fall outside the validity of the basic assumption should be checked, for example higher values indicating possible cementation or ageing.

5.5 Correlations for Sands

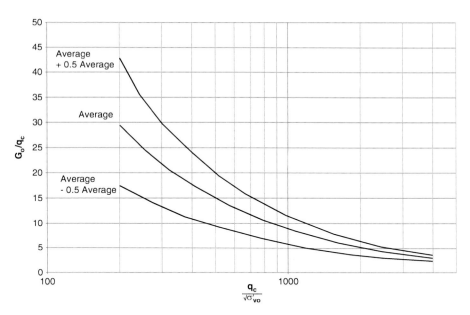

Fig. 5.8 G_o/q_c (Adapted from Rix and Stokes 1992)

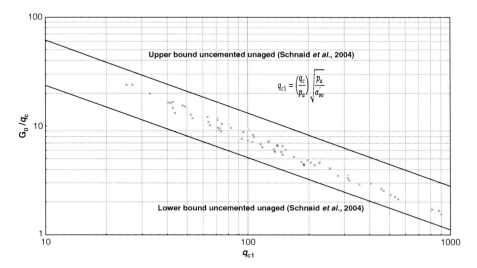

Fig. 5.9 Theoretical correlation between G_o/q_c and q_{c1} (Adapted from Schnaid 2009)

5.6 Correlations for Cohesive Soils

5.6.1 Correlation with Undrained Shear Strength of Cohesive Soils

Undrained shear strength of a clayey soil is one of the most derived parameters from a CPT test, apart from classification of the soil profile. The CPT test cannot directly measure the shear strength and only could be derived by empirical correlations associated with other tests, either in situ or laboratory. It is universally known and highlighted elsewhere in this book that the measured undrained shear strength is not unique and that it depends on many factors including rate of strain, loading arrangement, anisotropy, stress history etc. Therefore, when the CPT test result is calibrated against a particular test whether it is the vane test, direct shear or triaxial compression, the resulting values would be related to that particular test.

The commonly adopted test to calibrate against CPT is the vane shear test although other tests such as triaxial tests are also used.

Extensive research and studies have been conducted over the years and various theories postulated. However, the common, standard expression adopted by practitioners to derive c_u is based on Terzaghi's bearing capacity equation and can be written as follows:

$$c_u = \frac{(q_c - \sigma_{vo})}{N_c} \qquad (5.30)$$

where N_c is called the cone factor.

With the emergence of CPTu and the measurement of pore pressures, as previously discussed, the tip resistance q_c measured by CPTu needs to be corrected to yield q_t (see Eq. (5.3)), and Eq. (5.30) could be re-written with a different cone factor N_{kt}.

$$c_u = \frac{(q_t - \sigma_{vo})}{N_{kt}} \qquad (5.31)$$

Many studies have predicted N_{kt} values in the region of 10–20, sometimes even outside this range, and therefore a universally accepted unique value is not possible. It is strongly recommended that additional tests such as vane shear, direct shear, triaxial tests be conducted at test locations adjacent to cone penetrometer tests and calibrate to establish site specific N_{kt} values. Where such luxury is not available and there is no previous experience in the particular soil deposit, practitioners tend to adopt an N_{kt} value in the range of 14–16. However, it is recommended that sensitivity analysis be carried out especially where it is critical to the design.

Several researchers have attempted to identify correlations between the cone factor and the plasticity of the material. Aas et al. (1986) concluded that N_{kt} increases with plasticity although there was significant scatter in the results. Discussion by Powell and Quarterman (1988) and recent research by Kim et al. (2010) appear to confirm the conclusions of Aas et al. Robertson and Cabal (2014) also state that N_{kt} tends to increase with plasticity and decrease with sensitivity. However, others have found either no correlation or, in fact, N_{kt} decreasing with increasing plasticity index (La Rochelle et al. 1988; Hong et al. 2010; Remai 2013). Kulhawy and Mayne (1990) mention the work by Battaglio et al. (1973) who concluded that a trend exists for N_k for uncorrected vane shear test data in terms of plasticity index (*PI*). However, later reanalysis after correcting the vane hear test data indicated that N_k is not influenced by *PI*. Until further research confirms a correlation it is advised not to make any judgment unless site specific data is available.

It is suspected that the cone resistance is subjected to errors especially in softer clays. Therefore, researchers have proposed an alternative to derive c_u from excess pore pressure measurements. The corresponding equation could be written as follows:

$$c_u = \frac{\Delta u}{N_{\Delta u}} \quad (5.32)$$

where

$\Delta u = u_2 - u_0 =$ excess pore pressure measured at u_2 position and
$N_{\Delta u} =$ Pore pressure cone factor

Equations (5.7), (5.31) and (5.32) lead to the following relationship between $N_{\Delta u}$ and N_{kt}:

$$N_{\Delta u} = B_q N_{kt} \quad (5.33)$$

Robertson and Cabal (2014) indicate that $N_{\Delta u}$ varies between 4 and 10. Findings of other researchers found values of similar order, perhaps slightly narrower range (La Rochelle et al. 1988).

5.6.2 Correlation with Sensitivity of Cohesive Soils

Sensitivity can be defined as follows:

$$\text{Sensitivity}(S_t) = \frac{\text{peak undisturbed shear strength}}{\text{remoulded shear strength}}$$

One would deduce that sensitivity is somewhat related to the skin friction measured by the cone penetrometer. Schmertmann (1978) proposal was based on this and could be simply written as follows:

$$S_t = \frac{N_S}{R_f} \tag{5.34}$$

where N_s is a constant and R_f is the friction ratio.

Although Schmertmann (1978) suggested values for N_s it was based on a mechanical cone and therefore not applicable to electric cones including piezocones. While several proposals have been put forward, Lunne et al. (1997) suggestion of an average value of 5 with 6–9 likely range appears practical and reasonable.

5.6.3 Correlation with Over Consolidation Ratio of Cohesive Soils

Over consolidation ratio is an important parameter for cohesive soil deposits and is expressed as the ratio of the maximum past effective stress a soil element had ever been subjected to and the current effective vertical stress.

$$Over\,consolidation\,ratio\,(OCR) = \frac{maximum\,past\,effective\,pressure}{current\,effective\,vertical\,stress} = \frac{\sigma'_p}{\sigma'_v} \tag{5.35}$$

There are several methods to derive OCR from cone penetrometer data and some of the common or simple methods are given below.

Method 1 (Based on Mayne 2007).
- Estimate the overburden pressure σ_v
- Calculate σ'_p using Eq. 5.36

$$\sigma'_p = k(q_t - \sigma_v) \tag{5.36}$$

where k is a preconsolidation cone factor. An average of 0.33 is proposed with an expected range of 0.2–0.5
- Assess σ'_v and calculate OCR

Mayne (2007) points out that this is a first order estimate of the OCR for intact clays; it underestimates values for fissured clays. A similar order of factor k is recommended by Demers and Leroueil (2002)

Method 2
- Estimate σ'_v
- Estimate c_u from CPT or CPTu as discussed previously in this Section
- Calculate c_u/σ'_v (i.e. $(c_u/\sigma'_v)_{OC}$)
- Adopt $(c_u/\sigma'_p)_{NC}$ (say 0.22 as per Eq. (8.16)

5.6 Correlations for Cohesive Soils

Table 5.7 Constrained modulus coefficient for cohesive soils (Adopted from Sanglerat 1972; after Frank and Magnan 1995)

Soil type	q_c (MPa)	α
Low Plasticity Clay (CL)	$q_c < 0.7$	$3 < \alpha < 8$
	$0.7 < q_c < 2$	$1 < \alpha < 5$
	$q_c > 2$	$1 < \alpha < 2.5$
Silt, Low Liquid Limit (ML)	$q_c < 2$	$3 < \alpha < 6$
	$q_c > 2$	$1 < \alpha < 2$
High Plasticity Clay (CH)/ Silt, High Liquid Limit (MH)	$q_c < 2$	$2 < \alpha < 6$
	$q_c > 2$	$1 < \alpha < 2$
Organic Silt (OL)	$q_c < 1.2$	$2 < \alpha < 8$
Organic Clay (OH)/Peat (Pt)	$q_c < 0.7$	
$50 < w < 100$		$1.5 < \alpha < 4$
$100 < w < 200$		$1 < \alpha < 1.5$
$w > 300$		$\alpha < 0.4$

Note – w = natural moisture content (%)

- Calculate *OCR* using Eq. (5.37)

$$\left(\frac{c_u}{\sigma_v'}\right)_{OC} = \left(\frac{c_u}{\sigma_v'}\right)_{NC} OCR^m \qquad (5.37)$$

where m could be assumed to be 0.8 (see Sect. 8.6).

Method 3 (Based on Mayne and Kemper 1988)

$$OCR = 0.37 \left(\frac{q_c - \sigma_o}{\sigma_o'}\right)^{1.01} \qquad (5.38)$$

5.6.4 Correlation with Constrained Modulus of Cohesive Soils

Constrained modulus of soft to firm clays is commonly assessed using the consolidation test:

$$M = 1/m_v \qquad (5.39)$$

where m_v is the coefficient of volume compressibility.

Table 5.7 and Eq. (5.40) from Frank and Magnan (1995), who cites the work of Sanglerat (1972), could be used to obtaine the constrained modulus. Equation (5.40) is similar in nature to Eq. (5.23) for sands.

Table 5.8 Constrained modulus factor for clays

Author	α_M
Meigh (1987)	2–8
Kulhawy and Mayne (1990)	8.25
Mayne (2001)	8
Robertson (2009)	Q_t for Q_t <14 when I_c >2.2
	14 for Q_t >14 when I_c >2.2
	$0.03\left[10^{(0.55 I_c + 1.68)}\right]$ when I_c <2.2

Table 5.9 Coefficient of constrained modulus factor for NC and lightly OC clays and silts (After Meigh (1987))

Soil	Classification	α_M
Highly plastic clays and silts	CH, MH	2.5–7.5
Clays of Intermediate Plasticty	CI, CL	
q_c<0.7MN/m^2		3.7–10
q_c>0.7MN/m^2		2.5–6.3
Silts of Intermediate or low Plasticity	MI, ML	3.5–7.5
Organis silts	OL	2.5–10
Peat and organic clay	Pt, OH	
50<w<100 %		1.9–5
100<w<200 %		1.25–1.9
w>200 %		0.5–1.25

$$M = \alpha q_c \tag{5.40}$$

More recent studies indicate that M could be obtained from CPT profile using Eq. (5.41) which is of a similar form to Eq. (5.25) for sands.

$$M = \alpha_M (q_t - \sigma_{vo}) \tag{5.41}$$

Some of the values for α_M reported in the literature are shown in Table 5.8.

A more detailed assessment for different materials was carried out by Sanglerat (1972). However, as Meigh (1987) points out the recommendations were based on a Dutch Cone penetrometer. Meigh (1987), based on limited field evidence, adopted a factor (up) of 1.25 to the values proposed by Sanglerat (1972) and the modified factors are shown in Table 5.9.

5.6.5 Correlation with Compressibility of Cohesive Soils

Centre for Civil Engineering Research and Codes (CUR) (1996) suggests the following formula to calculate the compression ratio, CR:

5.6 Correlations for Cohesive Soils

Table 5.10 Coefficient β in Eq.(5.42) (After CUR 1996)

Soil type	Coefficient β
Sandy clay	0.2–0.4
Pure clay	0.4–0.8
peat	0.8–1.6

$$\frac{1}{CR} = \frac{q_c}{2.3\beta\sigma'_v} \tag{5.42}$$

The coefficient β for different soil types of cohesive materials is presented in Table 5.10.

5.6.6 Correlation with Friction Angle of Cohesive Soils

For soft to firm clays where c' can be assumed to be zero, Eq. (5.43) could be used to find ϕ' where it is related to B_q (Mayne 2014). The assumption of $c'=0$ preclude the use of this equation to assess the friction angle of OC soils. Mayne (2014) states that the equation is only applicable if $20° < \phi' < 45°$ and $0.1 < B_q < 1.0$.

$$\phi'\ (degrees) = 29.5 B_q^{0.121} \left[0.256 + 0.336 B_q + \log Q_t\right] \tag{5.43}$$

Where $Q_t = \frac{q_t - \sigma_{vo}}{\sigma'_{vo}}$ as previously defined.

5.6.7 Correlation with Small Strain Shear Modulus of Cohesive Soils

As previously discussed, the small strain shear modulus is related to the shear wave velocity, v_s. Mayne and Rix (1995) presented the following equation to obtained v_s from q_c values of intact and fissured clays.

$$v_s = 1.75\ (q_c)^{0.627} \tag{5.44}$$

where v_s is in units of m/s and q_c in kPa.

By considering as a bearing capacity problem, Mayne and Rix (1993) stated that it is more appropriate to utilize a net cone resistance, such as $(q_c - \sigma_{vo})$, or more correctly $(q_t - \sigma_{vo})$. The data for both fissured and intact clays were re-examined but no statistical improvement could be detected. Therefore only the intact clay results were re-examined and the following correlation resulted:

Table 5.11 CPT – V_s Correlation equations (After Wair et al. 2012)

Soil Type	Study	Geological Age	V_s(m/s)
All soils	Hegazy and Mayne (1995)	Quaternary	$\{10.1\log(q_c) - 11.4\}^{1.67}\left(100 f_s/q_c\right)^{0.3}$
	Mayne (2006)	Quaternary	$118.8\log(f_s) + 18.5$
	Piratheepan (2002)	Holocene	$32.3\, q_c^{0.089}\, f_s^{0.121}\, D^{0.215}$
	Andrus et al. (2007)	Holocene & Pleistocene	$2.62\, q_t^{0.395}\, I_c^{0.912}\, D^{0.124}\, SF^a$
	Robertson (2009)	Quaternary	$\left\{\left(10^{(0.55 I_c + 1.68)}\right)(q_t - \sigma_v)/p_a\right\}^{0.5}$
Sand	Sykora and Stokoe (1983)	–	$134.1 + 0.0052\, q_c$
	Baldi et al. (1989)	Holocene	$17.48\, q_c^{0.33}\, \sigma_v'^{0.27}$
	Hegazy and Mayne (1995)	Quaternary	$13.18\, q_c^{0.192}\, \sigma_v'^{0.179}$
	Hegazy and Mayne (1995)	Quaternary	$12.02\, q_c^{0.319}\, f_s^{-0.0466}$
	Piratheepan (2002)	Holocene	$25.3\, q_c^{0.163}\, f_s^{0.029}\, D^{0.155}$
Clay	Hegazy and Mayne (1995)	Quaternary	$14.13\, q_c^{0.359}\, e_0^{-0.473}$
	Hegazy and Mayne (1995)	Quaternary	$3.18\, q_c^{0.549}\, f_s^{0.025}$
	Mayne and Rix (1995)	Quaternary	$9.44\, q_c^{0.435}\, e_0^{-0.532}$
	Mayne and Rix (1995)	Quaternary	$1.75\, q_c^{0.627}$
	Piratheepan (2002)	Quaternary	$11.9\, q_c^{0.269}\, f_s^{0.108}\, D^{0.127}$

$p_a = 100$ kPa ; [a]$SF = 0.92$ for Holocene and 1.12 for Pleistocene
Stress unit in kPa and depth (D) in meters

$$v_s = 9.44\, (q_c)^{0.435}\, (e_0)^{-0.532} \tag{5.45}$$

where v_s is in m/s and q_c in kPa.

Mayne and Rix (1993) developed the following equation for clays where G_{max} is the initial tangent shear modulus (i.e. same as G_0)

$$G_{max} = 99.5\, p_a^{0.305}\, (q_c)^{\frac{0.695}{e_0^{1.130}}} \tag{5.46}$$

where p_a, G_{max} and q_c of same units and e_0 is the initial void ratio.

Various other authors have provided correlations for v_s and q_c. Wair et al. (2012) summarized v_s prediction equations as shown in Table 5.11. For consistency, Wair et al. (2012) have modified the equations to use consistent units of kPa for q_c, f_s and σ_v' with depth in metres.

5.7 Correlation with Unit Weight

There have been several studies linking the unit weight to CPT measurements. There have also been correlations for the unit weight involving the shear wave velocity. Two of the recent correlations for unit weight derivation based on measurements of CPT are by Robertson and Cabal (2010) and Mayne et al. (2010) who have used the sleeve friction in addition to the tip resistance to propose useful relationships. Robertson and Cabal (2010) proposal is shown in Fig. 5.10 and in Eq. (5.56) while Mayne et al. (2010) derivation is shown in Eq. (5.47).

$$\frac{\gamma}{\gamma_w} = 0.27\left[logR_f\right] + 0.36\left[log(q_t/p_a)\right] + 1.236 \qquad (5.47)$$

$$\gamma_t = 1.95\gamma_w\left(\frac{\sigma'_{vo}}{p_a}\right)^{0.06}\left(\frac{f_s}{p_a}\right)^{0.06} \qquad (5.48)$$

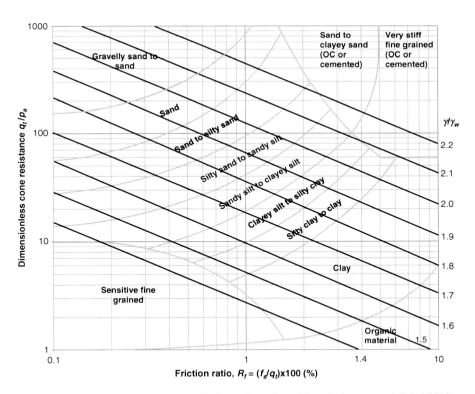

Fig. 5.10 Normalised unit weight and friction ratio (Adapted from Robertson and Cabal 2010)

Table 5.12 Permeability from CPT results

Soil Behaviour Type	Soil Permeability (m/s)
Sensitive fine grained	3×10^{-9} to 3×10^{-8}
Organic soils-peats	1×10^{-8} to 1×10^{-6}
Clays-clay to silty clay	1×10^{-10} to 1×10^{-7}
Silt mixtures clayey silt to silty clay	3×10^{-9} to 1×10^{-7}
Sand mixtures; silty sand to sandy silt	1×10^{-7} to 1×10^{-5}
Sands; clean sands to silty sands	1×10^{-5} to 1×10^{-3}
Gravelly sand to sand	1×10^{-3} to 1
[a]Very stiff sand to clayey sand	1×10^{-8} to 1×10^{-6}
[a]Very stiff fine grained	1×10^{-9} to 1×10^{-7}

[a]Over consolidated or cemented

5.8 Correlation with Permeability

Robertson (1990) provided a correlation between the Soil Behaviour Type (see Sect. 5.4) and the permeability as shown in Table 5.12. Robertson and Cabal (2010) state that the average permeability (k) shown in Table 5.12 can be represented by Eq. (5.49) and Eq. (5.50) using SBT index, I_c, defined by Eq. (5.11).

When $\quad 1.0 < I_c \leq 3.27 \quad k = 10^{(0.952 - 3.04 \, I_c)} \, m/s \quad (5.49)$

When $\quad 3.27 < I_c \leq 4.0 \quad k = 10^{(-4.52 - 1.37 \, I_c)} \, m/s \quad (5.50)$

5.9 Correlation with SPT N

Standard penetration test (SPT) is probably the most used in situ device worldwide to assess the soil characteristics in situ. Some designers prefer to convert CPT results to equivalent SPT N values if the design methods are based on SPT N value, and vice versa.

As the CPT is a continuous test it provides more details than SPT test results because of the nature of the test being continuous and therefore detects local changes in the profile. This is most evident in soft to firm soils where SPT N values are generally found to be zero or a very low value. Such results do not provide a great deal of information to the designer except perhaps that the clay is very soft or soft. Conversely, the CPT provides significantly more details which could be used to interpret standard parameters. This is evident in the example shown in Fig. 5.11a which shows a cluster of CPT results converted to c_u values. In the same Figure, available vane shear test results, some of which used to calibrate CPT results, are also plotted. Figure 5.11b shows the available SPT N values from boreholes located in the same area. It is evident that the information available for the weak layers from the SPT results is significantly less useful to the designer than CPT plots. For stiffer soils and sandy soils, SPT N values provide much more useful information and the reason its use worldwide in routine foundation designs.

5.9 Correlation with SPT N

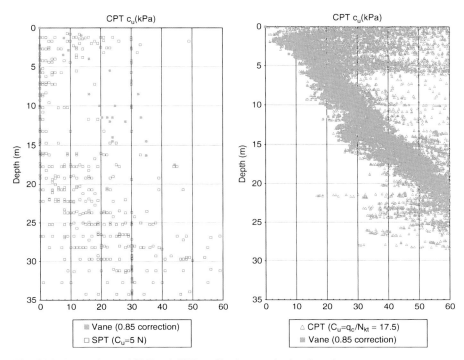

Fig. 5.11 Comparison of SPT and CPT profiles in a weak clay deposit

Table 5.13 Ratios of q_c/N (After Sanglerat 1972; Schmertmann 1970, 1978)

Soil	q_c (kPa)/N_{60}
Silts, sandy silts, slightly cohesive silt-sand mix	200[a] (200–400)[b]
Clean fine to medium sands and slightly silty sands	300–400[a] (300–500)[b]
Coarse sands and sands with little gravel	500–600[a] (400–500)[b]
Sandy gravel and gravel	800–1000[a] (600–800)[b]

[a]Values proposed by Sanglerat (1972) and reported in Peck et al. (1974)
[b]Values suggested by Schmertmann (1970, 1978), reported by Holtz (1991) in parentheses

When results of both types of tests are not available CPT-SPT relationships become very useful to the designer. Various authors have provided CPT-SPT relationships mostly based on experienced gained from in situ testing and calibrating against each other.

Sivakugan and Das (2010) summarized the work of Sanglerat (1972) and Schmertmann (1970, 1978) on the ratio of q_c/N for different soils, as shown in Table 5.13. The results indicate that the ratio is low, 200–400 (kPa), for fine grained materials, increasing to 800–1000 (kPa) for gravels.

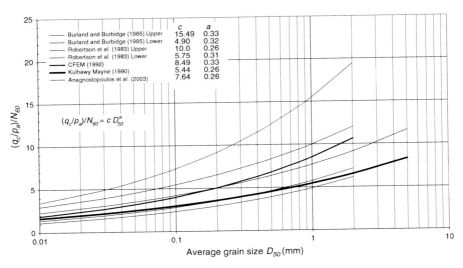

Fig. 5.12 q_c/N_{60} variation with grain size (Adapted from Robertson and Campanella 1983)

As Robertson and Campanella (1983) investigated they found that q_c/N ratio values published had a very wide scatter. However they found that the results could be rationalized somewhat when the ratio is pitted against the mean grain size, D50. (see Fig. 5.12).

The importance of the grain size in penetration test results is evident from Fig. 5.12, the plot of the variation of q_c/N with the median grain size D_{50}, and the upper and lower bounds, from Robertson et al. (1983). In the same Figure, data from Burland and Burbidge (1985), Canadian Foundation Engineering Manual (Canadian Geotechnical Society 1992), Kulhawy and Mayne (1990) and Anagnostopoulos et al. (2003) are also shown, some as lower and upper bounds or average values. The curves shown in Fig. 5.12 take the form of:

$$\frac{\left(\frac{q_c}{p_a}\right)}{N_{60}} = c D_{50}{}^a \tag{5.51}$$

where the values of 'a' and 'c' are shown in Fig. 5.12.

Some of these relationships are based on a significant data base and/or relevant to a particular locality and this is one of the reasons for the differences in the coefficients 'a' and 'c'. For example, Kulhawy and Mayne (1990) equation has the coefficients $c = 5.44$ and $a = 0.26$:

$$\frac{\left(\frac{q_c}{p_a}\right)}{N_{60}} = 5.44 D_{50}^{0.26} \tag{5.52}$$

However, the proposed relationship by Anagnostopoulos et al. (2003) for Greek soils has $c = 7.64$ although a remains the same:

$$\frac{\left(\frac{q_c}{p_a}\right)}{N_{60}} = 7.64 D_{50}^{0.26} \tag{5.53}$$

To use the above relationships the median grain size (D_{50}) is required. If test results are not available, D_{50} needs to be estimated. Jefferies and Davies (1993) proposed a method for the estimation of N_{60} values directly from CPTu results without resorting to soil sampling and laboratory testing. They proposed the use of soil behavior type (SBT) index (see Sect. 5.4) to correlate CPT and SPT. Lunne et al. (1997) applied this technique to the SBT Index, I_c, defined in Eq. (5.11) to give the following relationship.

$$\frac{q_c/p_a}{N_{60}} = 8.5 \left(1 - \frac{I_c}{4.6}\right) \tag{5.54}$$

where I_c = Soil Behaviour Type in Eq.(5.11)

Jefferies and Davies (1993) suggest that the CPTu has a fivefold improved precision compared to the SPT. Robertson and Cabal (2014) agree with Jeffries and Davies (1993) that the most reliable way to obtain SPT N values is to perform a CPT and convert the CPT to an equivalent SPT.

5.10 Correlation with Bearing Capacity

5.10.1 Shallow Foundations

De Court (1995) proposed the following correlation with tip resistance, q_c, from electric CPT to calculate the ultimate bearing capacity of a shallow foundation for sand:

$$\text{Ultimate bearing capacity} = q_{ult} = \frac{q_c}{4} \tag{5.55}$$

Frank and Magnan (1995) citing the French experience state the bearing capacity of shallow foundations according to MELT (1993) is as follows:

$$q_{ult} = k_c q_c + q_o \tag{5.56}$$

where k_c is the bearing factor given in Table 5.14.

Table 5.14 Bearing capacity factor k_c (CPT) for shallow foundations (After MELT 1993)

Soil type		Expression for k_c
Clay/Silt	Soft to Hard	$0.32 \left[1 + 0.35\left(0.6 + 0.4\frac{B}{L}\right)\frac{D}{B}\right]$
Sand/Gravel	Loose	$0.14 \left[1 + 0.35\left(0.6 + 0.4\frac{B}{L}\right)\frac{D}{B}\right]$
	Medium	$0.11 \left[1 + 0.50\left(0.6 + 0.4\frac{B}{L}\right)\frac{D}{B}\right]$
	Dense	$0.08 \left[1 + 0.85\left(0.6 + 0.4\frac{B}{L}\right)\frac{D}{B}\right]$
Chalk	Weathered	$0.17 \left[1 + 0.27\left(0.6 + 0.4\frac{B}{L}\right)\frac{D}{B}\right]$

B = width; L = length; D = embedment

Table 5.15 Base bearing capacity factor k_c (CPT) for deep foundations (After MELT 1993)

Soil type		q_c (MPa)	k_c ND	k_c D
Clay/Silt	soft	<3	0.40	0.55
	stiff	3–6		
	hard (clay)	>6		
Sand/Gravel	loose	<5	0.15	0.50
	medium	8–15		
	dense	>20		
Chalk	soft	<5	0.20	0.30
	weathered	>5	0.30	0.45

ND = non displacement pile, D = displacement pile

5.10.2 Deep Foundations

Frank and Magnan (1995) in their national report for France for the CPT95 conference state that the French practice does not use the sleeve friction f_s to calculate bearing capacity but only the cone resistance q_c. The ultimate base resistance is given by:

$$q_b = k_c q_c \tag{5.57}$$

where k_c is the base bearing capacity factor.

The values of k_c recommended by the design code for foundations of the French road administration (MELT 1993) are given in Table 5.15. The unit skin friction q_s is obtained from using the following equation.

$$q_s = \text{minimum value of } \{q_c/\beta; q_{smax}\} \tag{5.58}$$

where β and q_{smax} values recommeneded for different soil types and pile types are given in Table 5.16. Frank and Magnan (1995) state that CPT pile design rules used in France are based on a large database of full scale pile loading tests and therefore probably one of the best design rules used.

5.10 Correlation with Bearing Capacity

Table 5.16 Unit limit skin friction from CPT (After MELT 1993)

Pile type		Clay and silt			Sand and gravel			Chalk	
		Soft	Stiff	Hard	Loose	Medium	Dense	Soft	Weathered
Drilled	β	–	– / 75[a]	– / 80[a]	200	200	200	125	80
	q_{smax} (kPa)	15	40	40	–	–	120	40	120
Drilled removed casing	β	–	100 / 100[b]	– / 100[b]	250	250	300	125	100
	q_{smax} (kPa)	15	40 / 60[b]	40 / 80[b]	–	40	120	40	80
Steel driven closed-ended	β	–	120	150	300	300	300	[c]	
	q_{smax} (kPa)	15	40	80	–	–	120		
Driven concrete	β	–	75	–	150	150	150	[c]	
	q_{smax} (kPa)	15	80	80	–	–	120		

[a] Trimmed and grooved at the end of drilling
[b] dry excavation, no rotation of casing
[c] in chalk, q_s can be very low for some types of piles; a specific study is needed

5.11 Liquefaction Assessment

As discussed in Sect. 4.11 in Chap. 4, soil liquefaction is a serious phenomenon which could create major destruction to life and property during earthquakes. The use of in situ tests to characterize soil liquefaction is quite popular and it is widely accepted that they provide better solutions than laboratory tests because of inherent difficulties associated with collecting undisturbed samples from sandy soils. Historically, SPT test results were used to assess liquefaction potential, however, with the advent and rapid advancement of the CPT, the latter has provided an additional tool that could be used for the same purpose. Table 5.17 lists the advantages and disadvantages of the two methods.

CPT is now accepted as a primary tool for liquefaction assessment because of, as Robertson and Campanella (1995) point out, simplicity, repeatability and accuracy. CPT also provides a continuous record in addition to being quicker and less costly. It is quite common to use both SPTs and CPTs especially when the project is large and/or critical.

Liquefaction potential is generally assessed based on a factor of safety related to the soil resistance. The factor of safety (FS_{li}) is defined as the ratio of cyclic shear strength of the soil (i.e., cyclic resistance ratio, CRR) and the cyclic stress developed by the design earthquake (i.e., cyclic stress ratio, CSR):

$$FS_{li} = CRR/CSR \tag{5.59}$$

The use of appropriate FS_{li} depends on the site, information available, assessment tools and design assumptions made. If design assumptions are conservative and good quality data are available and the designer adopts a conservative approach a FS_{li} of 1 could be adopted if valid computational method is used. Even then, a higher value may be warranted if the consequences of failure could have a significant effect on the environment and health and safety.

Table 5.17 Comparison of SPT and CPT for assessment of liquefaction potential (After Youd et al. 2001)

Feature	SPT	CPT
Number of test measurements at liquefaction sites	Abundant	Abundant
Type of stress-strain behavior influencing test	Partially drained, large strain	Drained, large strain
Quality control and repeatability	Poor to good	Very good
Detection variability of soil deposits	Good	Very good
Soil types in which test is recommended	Non-gravel	Non-gravel
Test provides sample of soil	Yes	No
Test measures index or engineering property	Index	Index

5.11 Liquefaction Assessment

CRR and CSR values are often adjusted to an equivalent shear stress induced by an earthquake magnitude $M = 7.5$ and commonly referred to as $CRR_{7.5}$ and $CSR_{7.5}$ and therefore Eq. (5.59) could be written as follows:

$$FS_{li} = {}^{CRR_{7.5}}\big/{}_{CSR_{7.5}} \qquad (5.60)$$

The methodology to assess liquefaction potential or FS_{li} could be summarized as follows (Idriss and Boulanger 2006):

1. Calculate CSR for the design earthquake, M (CSR_M) – Sect. 5.11.1
2. Calculate Magnitude Scaling Factor (MSF) – Sect. 5.11.1.1
3. Convert CSR_M to a standard earthquake magnitude of 7.5 (i.e. $CSR_{7.5}$) – Sect. 5.11.1
4. Normalise resistance and correct for overburden stress – Sect. 5.11.2
   (a) CPT– Sect. 5.11.2.1
   (b) SPT – Sect. 5.11.2.2
5. Correct for Fines Content, FC – Sect. 5.11.2.3
6. Calculate CRR for a magnitude of 7.5 and 1 atmosphere, i.e. $CRR_{7.5,100}$ – Sect. 5.11.3
7. Calculate CRR – Sect. 5.11.3 (after assessing the overburden factor, K_σ – Sect. 5.11.3.1)
8. Calculate FS_{li} – Sect. 5.11 introduction

5.11.1 Cyclic Stress Ratio

Cyclic Stress Ratio (CSR) for a design earthquake magnitude M could be calculated by the following (Seed and Idriss 1971):

$$CSR_M = 0.65 \left(\frac{a_{max}}{g}\right)\left(\frac{\sigma_{vo}}{\sigma'_{vo}}\right) r_d \qquad (5.61)$$

where

a_{max} maximum horizontal acceleration at the ground surface
g acceleration due to gravity
σ_{vo} total vertical overburden stress
σ'_{vo} effective overburden stress
r_d stress reduction factor

Idriss (1999) proposed the following expressions to compute the stress reduction factor, r_d:

$$r_d = e^{\left(\alpha_{(z)} + \beta_{(z)} M\right)} \tag{5.62a}$$

$$\alpha_{(z)} = -1.012 - 1.126 \, Sin\left(\frac{z}{11.73} + 5.133\right) \tag{5.62b}$$

$$\beta_{(z)} = 0.106 - 0.118 \, Sin\left(\frac{z}{11.28} + 5.142\right) \tag{5.62c}$$

$z =$ depth below ground surface ($z \leq 34$m)

As Idriss and Boulanger (2008) state, although the above equations are mathematically applicable to a depth of $z \leq 34$ m, uncertainty in r_d increases with depth and therefore should only be applied to depths less than 20 m or so. For deeper sites specific site response analysis should be carried out.

Other correlations have been proposed and the following tri-linear function provides a good fit to r_d originally proposed by Seed and Idriss in 1971 (Youd et al. 2001):

$$r_d = 1.0 - 0.00765 \, z \; if \; z \leq 9.15 \, m \tag{5.63a}$$
$$r_d = 1.174 - 0.0267 \, z \; if \; 9.15 \, m < z \leq 23 \, m \tag{5.63b}$$

Magnitude Scaling Factor (*MSF*) is used to convert the *CSR* calculated for the design earthquake to a common or reference earthquake magnitude, generally accepted to be of magnitude 7.5 using Eq. (5.64).

$$CSR_{7.5} = CSR_M / MSF \tag{5.64}$$

5.11.1.1 Magnitude Scaling Factor (*MSF*)

There are several expressions that could be used to calculate *MSF*. The participants at the NCEER Workshops in 1996/98 recommended the following expression as a lower bound for *MSF* (Youd et al. 2001):

$$MSF = {10^{2.24}}/{M^{2.56}} \tag{5.65}$$

The MSF relationship was re-evaluated by Idriss (1999) as reported by Idriss and Boulanger (2006) and presented below.

$$MSF = 6.9 \, e^{\left(\frac{-M}{4}\right)} - 0.058 \leq 1.8 \tag{5.66}$$

5.11.2 *Normalization of Resistance*

The methodology described below is as described by Boulanger (2003) and, Idriss and Boulanger (2006) including fines correction proposed by Robertson and Wride (1998). To be consistent and simple, it is assumed that the unit for pressure/stress would be kPa.

5.11.2.1 Normalization of Resistance – CPT

CPT penetration resistance, q_c, is initially corrected for overburden stress effects using an equivalent effective vertical stress (σ'_{vo}) of one atmosphere (100 kPa) and an overburden correction factor, C_N, as part of the semi-empirical procedure (Boulanger 2003):

$$q_{cN} = \frac{q_c}{100} \tag{5.67}$$

$$q_{c1N} = C_N q_{cN} = C_N \frac{q_c}{100} \leq 254 \tag{5.68}$$

$$C_N = \left(\frac{100}{\sigma'_{vo}}\right)^\beta \leq 1.7 \tag{5.69}$$

$$\beta = 1,338 - 0.249 \left(q_{c1N}\right)^{0.264} \tag{5.70}$$

Solving for C_N requires an iterative process because of its dependence on q_{c1N}.

5.11.2.2 Normalization of Resistance – SPT

The equivalent equations for SPT are as follows (Boulanger 2003):

$$(N_1)_{60} = C_N (N)_{60} \tag{5.71}$$

$$C_N = \left(\frac{100}{\sigma'_{vc}}\right)^\alpha \leq 1.7 \tag{5.72a}$$

$$\alpha = 0.784 - 0.0768 \left\{(N_1)_{60}\right\}^{0.5} \tag{5.72b}$$

where σ'_{vc} is the operating effective vertical stress in kPa.

Solving for C_N requires an iterative process because of its dependence on $(N_1)_{60}$.

5.11.2.3 Correction for Fines Content

The above equations in Sects. 5.11.2.1 and 5.11.2.2 are based on the assumption that sands encountered are from clean sand deposits. It is generally accepted that correlations to obtain CRR values would be different if the sands have fines. Therefore a correction is made to adjust the SPT or CPT resistance to an equivalent clean sand value.

For the SPT test, the following relationship was proposed by Idriss and Boulanger 2004 for non-plastic sands:

$$\Delta(N_1)_{60} = \exp\left\{1.63 + \frac{9.7}{FC + 0.01} - \left(\frac{15.7}{FC + 0.01}\right)^2\right\} \quad (5.73a)$$

$$(N_1)_{60cs} = (N_1)_{60} + \Delta(N_1)_{60} \quad (5.73b)$$

In the case of CPT, Robertson and Wride (1997) and Suzuki et al. (1997) proposed the use soil behavior type index which is a function of q_c and F_r to obtain CRR when the fines content is high. However, as Idriss and Boulanger (2008) state, the curve proposed by Robertson and Wride (1997) is unconservative with similar comments on the proposal by Suzuki et al. (1997) for high fine contents. Idriss and Boulanger (2008) suggest the modification of cone resistance values to account for non-plastic fines in a similar way to the SPT corrections discussed above using the following relationships:

$$\Delta q_{c1N} = \left(5.4 + \frac{q_{c1N}}{16}\right)\exp\left\{1.63 + \frac{9.7}{FC + 0.01} - \left(\frac{15.7}{FC + 0.01}\right)^2\right\} \quad (5.74a)$$

where FC = Fines content

$$(q_{c1N})_{cs} = q_{c1N} + \Delta q_{c1N} \quad (5.74b)$$

5.11.3 Computation of Cyclic Resistance Ratio (CRR)

The Cyclic Resistance Ratio, CRR, could be obtained from the CPT and SPT results using Eqs. (5.75) and (5.76) respectively (Idriss and Boulanger 2004):

$$CRR_{7.5,\,100} = \exp\left\{\left(\frac{(q_{c1N})_{cs}}{540}\right) + \left(\frac{(q_{c1N})_{cs}}{67}\right)^2 - \left(\frac{(q_{c1N})_{cs}}{80}\right)^3 + \left(\frac{(q_{c1N})_{cs}}{114}\right)^4 - 3\right\} \quad (5.75)$$

$$CRR_{7.5,\,100} = \exp\left\{\left(\frac{(N_1)_{60CS}}{14.1}\right) + \left(\frac{(N_1)_{60CS}}{126}\right)^2 - \left(\frac{(N_1)_{60CS}}{23.6}\right)^3 + \left(\frac{(N_1)_{60CS}}{25.4}\right)^4 - 2.8\right\} \quad (5.76)$$

where $CRR_{7.5,100}$ relates to an earthquake magnitude of 7.5 and an effective stress of 100kPa.

The value of $CRR_{7.5,100}$ needs to be adjusted to the overburden to obtain $CRR_{7.5}$. This is carried out by the use of an overburden correction factor, K_σ (Boulanger 2003):

$$CRR_{7.5} = CRR_{7.5,100} K_\sigma \quad (5.77)$$

Section 5.11.3.1 describes the procedure to obtain the relevant K_σ value.

5.11.3.1 Assessment of Overburden Factor

Boulanger (2003) proposed the following expression for K_σ:

$$K_\sigma = 1 - C_\sigma \ln\left(\frac{\sigma'_{vc}}{100}\right) \leq 1.0 \qquad (5.78)$$

where C_σ could be obtained from the following expressions for the CPT and SPT (Boulanger 2003):

$$C_\sigma = \frac{1}{37.3 - 8.27 \left(q_{c1N}\right)^{0.264}} \leq 0.3 \qquad (5.79)$$

$$C_\sigma = \frac{1}{18.9 - 2.55 \left((N_1)_{60}\right)^{0.5}} \leq 0.3 \qquad (5.80)$$

References

Aas G, Lacasse S, Lunne T, Hoeg K (1986) Use of in situ tests for foundation design on clay. In: Proceedings of in situ '86: use of in situ tests in geotechnical engineering, Virginia, pp 1–30

Anagnostopoulos A, Koukis G, Sabatakakis N, Tsiambaos G (2003) Empirical correlation of soil parameters based on cone penetration tests (CPT) for Greek soils. Geotechn Geol Eng 21 (4):373–387

Andrus RD, Mohanan NP, Piratheepan P, Ellis BS, Holzer TL (2007) Predicting shear-wave velocity from cone penetration resistance. In: Proceedings, 4th international conference on earthquake geotechnical engineering, Thessaloniki, Greece

ASTM D4253-14 Standard test methods for maximum index density and unit weight of soils using a vibratory table

ASTM D4254-14 Standard test methods for minimum index density and unit weight of soils and calculation of relative density

Baldi, G, Bellotti R, Ghionna V, Jamiolkowski M, Pasqualini E (1986) Interpretation of CPT's and CPTU's. 2nd Part: Drained penetration. In: Proceedings 4th international geotechnical seminar, Singapore, pp 143–156

Baldi G, Bellotti R, Ghionna VN, Jamiolkowski M, Lo Presti, DFC (1989) Modulus of sands from CPTs and DMTs. In: Proceedings of the 12th international conference on soil mechanics and foundation engineering. Rio de Janeiro. Balkema Pub., Rotterdam, vol 1, pp 165–170

Battaglio M, Bruzzi D, Jamiolkowski M, Lancelotta R (1973) Interpretation of CPT and CPTU. In: Proceedings, Field Instrumentation on In-Situ Measurements, Singapore, pp 129-143

Battaglio M, Bruzzi D, Jamiolkowski M, Lancellotta, R (1986) Interpretation of CPT and CPTU. In: Interpretation of CPT and CPTU. In: Proceedings, field instrumentation on in-situ measurements, Singapore, pp 129–143

Bellotti R, Ghionna VN, Jamiolkowski M, Robertson PK (1989) Shear strength of sand from CPT. In: Proccedings, 12th international conference on soil mechanics and foundation engineering. Rio de Janeiro. Balkema Pub., Rotterdam, vol 1, pp 179–184

Bergdahl U, Ottosson E, Malmborg BS (1993) Plattgrundläggning. AB Svensk Byggtjänst, Stockholm

Boulanger RW (2003) State normalization of penetration resistance and the effect of overburden stress on liquefaction resistance. In: Proceedings 11th international conference on soil dynamics and earthquake engineering and 3rd international conference on earthquake geotechnical engineering, University of California, Berkeley

Burland JB, Burbidge MC (1985) Settlement of foundations on sand and gravel. In: Proceedings institute of civil engineers, Part I7, pp 1325–1381

Campanella RG, Robertson PK (1988) Current status of the piezocone test. In: Penetration testing 1988, vol 1 (Proc. ISOPT-1), Orlando, Balkema, Rotterdam, pp 93–116

Campanella RG, Gillespie D, Robertson PK (1982) Pore pressures during cone penetration testing. In: Proceedings of the 2nd european symposium on penetration testing, ESPOT II. Amsterdam. A.A. Balkema, pp 507–512

Canadian Geotechnical Society (1992) Canadian foundation engineering manual, 3rd edn. Bi-Tech Publishers, Ltd., Richmond

CUR (1996) Building on soft soils, Centre for Civil Engineering Research and Codes (CUR), CUR Report 162. A. A. Balkema, Rotterdam

Das BM, Sivakugan N (2011) Maximum and minimum void ratios and median grain size of granular soils: their importance and correlations with material properties. In: Proceedings of the international conference on advances in geotechnical engineering, ICAGE 2011, Perth, WA, Australia, pp 59–73

De Court L (1995) Prediction of load settlement relationships for foundations on the basis of the SPT-T. In: Ccico de Conferencias Intern. "Leonardo Zeevaert", UNAM, Mexico, pp 85–104

Demers D, Leroueil S (2002) Evaluation of preconsolidation pressure and the overconsolidation ratio from piezocone tests of clay deposits in quebec. Can Geotech J 39(1):174–192

Douglas BJ, Olsen RS (1981) Soil classification using electric cone penetrometer. In: Symposium on cone penetrometer testing and experience, Proceedings ASCE National convention, St Louis, pp 209–27, ASCE

Durgunoglu HT, Mitchell JK (1975) Static penetration resistance of soils, I-Analysis, II-Evaluation of theory and implications for practice. In: Proceedings of the conference on in situ measurement of soil properties, ASCE, Raleigh, North Carolina, vol I, pp. 151–189

Dysli M, Steiner W (2011) Correlations in soil mechanics. PPUR, EPFL Rolex Learning Centre, Lausanne

Eslaamizaad S, Robertson PK (1997) Evaluation of settlement of footings on sand from seismic in-situ tests. In: Proceedings of the 50th Canadian geotechnical conference, Ottawa, Ontario, vol 2, pp 755–764

Eslami A, Fellenius BH (1997) Pile capacity by direct CPT and CPTU data. In: Year 2000 Geotechnics, Geotechnical Engineering Conference, AIT Bangkok, 18 p

Frank R, Magnan JP (1995) Cone penetration testing in France: national report. In: Proceedings CPT '95, Linkoping, Swedish Geotechnical Society, vol 3, pp 147–156

Hegazy YA, Mayne PW (1995) Statistical correlations between VS and cone penetration data for different soil types. In: Proceedings international symposium on cone penetration testing, CPT '95, Linkoping, Sweden, vol 2, pp 173–178

Holtz RD (1991) Chapter 5: Stress distribution and settlement of shallow foundations. In: Fang H-Y (ed) Foundation engineering handbook, 2nd edn. van Nostrand Reinhold, New York, pp 166–222

Hong SJ, Lee MJ Kim JJ, Lee WJ (2010) Evaluation of undrained shear strength of Busan clay using CPT. In: 2nd international symposium on cone penetration testing CPT'10, pp 2–23

Idriss IM (1999) An update to the Seed-Idriss simplified procedure for evaluating liquefaction potential. In: Proceedings TRB workshop on new approaches to liquefaction, Publication No FHWA-RD-99-165, Federal Highway Administration

Idriss IM, Boulanger RW (2004) Semi-empirical procedures for evaluating liquefaction potential during earthquakes. In: Proceddings 11th international conference on soil dynamics and earthquake engineering, Berkeley, pp 32–56

Idriss IM, Boulanger RW (2006) Semi-empirical procedures for evaluating liquefaction potential during earthquakes. J Soil Dyn Earthquake Eng 26:115–130, Elsevier

Idriss IM, Boulanger RW (2008) Soil liquefaction during earthquakes. Earthquake Engineering Research Institute, Oakland

References

ISSMGE TC16 (1989) International reference test procedure for the CPT and CPTU. In: Proceedings 12th international conference on soil mechanics and geotechnical engineering, Amsterdam

Jamiolkowski M, Ladd CC, Germaine J, Lancellotta R (1985) New developments in field and lab testing of soils. In: Proceedings 11th international conference on soil mechanics and foundations engineering, vol 1, San Francisco, pp 57–154

Jamiolkowski M, LoPresti DCF, Manassero M (2001) Evaluation of relative density and shear strength of sands from cone penetration test and flat dilatometer test, Soil behaviour and soft ground construction (GSP 119). ASCE, Reston, pp 201–238

Janbu N (1963) Soil compressibility as determined by oedometer and triaxial tests. In: Proceedings European conference on soil mechanics and foundation engineering, Wiesbaden, 1, pp 19–25

Janbu N, Senneset K (1974) Effective stress interpretation of in situ static penetration tests. In: Proceedings European symposium on penetration testing, ISOPT, Stockholm, 2.2, pp 181–93

Jefferies MG, Davies MP (1991) Soil classification by the cone penetration test. Discussion. Can Geotech J 28:173–176

Jefferies MG, Davies MP (1993) Use of CPTU to estimate equivalent SPT N_{60}. Geotech Test J ASTM 16(4):458–468

Jones GA, Rust E (1982) Piezometer penetration testing. In: Proceedings 2nd European symposium on penetration testing, ESOPT-2, Amsterdam, 2, pp 697–614

Kim D, Shin Y, Siddiki N (2010) Geotechnical design based on CPT and PMT, Joint Transportation Research Program, FHWA/IN/JTRP-2010/07, Final Report

Kulhawy FH, Mayne PW (1990) Manual on estimating soil properties for foundation design. Report EL- 6800 submitted to Electric Power Research Institute, Palo Alto, California, 306 p

La Rochelle P, Zebdi PM, Leroueil S, Tavenas F, Virely D (1988) Piezocone tests in sensitive clays of eastern Canada. In: Proceedings international symposium on penetration testing, ISOPT-1, Orlando, 2, pp 831–41, Balkema Pub., Rotterdam

Long M (2008) Design parameters from insitu tests in soft ground – recent developments. In: Huang AB, Mayne PW (eds) Geotechnical and geophysical site characterization. Taylor & Francis Group, London

Lunne T, Christophersen HP (1983) Interpretation of cone penetrometer data for offshore sands. In: Proceedings of the offshore technology conference, Richardson, Texas, Paper No. 4464

Lunne T, Kleven A (1981) Role of CPT in North Sea Foundation Engineering. In: Symposium on cone penetration and testing and experience, ASCE, pp 49–75

Lunne T, Robertson PK, Powell JJM (1997) Cone penetration testing in geotechnical practice. Blackie Academic & Professional/Chapman-Hall Publishers, London, 312 p

Mayne PW (2001) Stress-strain-strength-flow parameters from enhanced in-situ tests. In: Proceedings international conference on in-situ measurement of soil properties & case histories (In-Situ 2001), Bali, Indonesia, pp 27–47

Mayne PW (2006) In-situ test calibrations for evaluating soil parameters. In: Phoon KS, Hight DW, Leroueil S, Tan TS (eds) Proceedings of the second international workshop on characterization and engineering properties of natural soils. Singapore

Mayne PW (2007) Cone penetration testing State-of-Practice, NCHRP Synthesis. Transportation Research Board Report Project 20–05, 118 pages

Mayne PW (2014) Interpretation of geotechnical parameters from seismic piezocone tests. In: Robertson PK, Cabal KI (eds) Proceedings 3rd international symposium on cone pentrometer testing, CPT'14, Las Vegas, pp 47–73

Mayne PW, Kemper JB (1988) Profiling OCR in stiff clays by CPT and SPT. Geotech Test J 11(2):139–147

Mayne PW, Rix JG (1993) G_{max}-q_c relationships for clays. Geotech Test J ASTM 16(1):54–60

Mayne PW, Rix GJ (1995) Correlations between shear wave velocity and cone tip resistance in natural clays. Soils Found 35(2):107–110

Mayne PW, Christopher BR, DeJong JT (2002) Manual on Subsurface Investigations, FHWA Publication No. FHWA NHI-01-031, p 294

Mayne PW, Peuchen J, Bouwmeester D (2010) Soil unit weight estimation from CPTS, In: Proceedings 2nd international symposium on cone penetration testing CPT '10, vol 2, Huntington Beach, California, pp 169–176

Meigh AC (1987) Cone penetration testing – methods and interpretation, CIRIA, Butterworths

MELT (1993) Régles techniques de conception et de calcul des fondations des ouvrages de génie civil, Fascicule No 62, Titre V No 93-3 TO. Ministére de l'Equpment du Logement et des Transports, Paris

Meyerhoff GG (1956) Penetration tests and bearing capacity of cohesionless soils. J Soil Mech Found Div ASCE 82(SM1):1–19

Molle J (2005) The accuracy of the interpretation of CPT-based soil classification methods in soft soils, MSc thesis, Section for Engineering Geology, Department of Applied Earth Sciences, Delft University of Technology, Report No. 242, Report AES/IG/05-25

Olsen RS, Mitchell JK (1995) CPT stress normailisation and prediction of soil classification. In: Proceedings international symposium on cone penetration testing, CPT'95, Lingköping, Sweden, Swedish Geotechnical Society, 2, pp 257–262

Peck RB, Hanson WE, Thornburn TH (1974) Foundation design. Wiley, New York

Piratheepan P (2002) Estimating shear-wave velocity from SPT and CPT Data, Master of Science thesis, Clemson University

Powell JJM, Quarterman RST (1988) The interpretation of cone penetration tests in clays, with particular reference to rate effects. In: Proceedings international symposium on penetration testing, ISPT-1, Orlando, 2, pp 903–910, Balkema Pub., Rotterdam

Ramsey N (2002) A calibrated model for the interpretation of cone penetration tests (CPT's) in North Sea quaternary soils. In: Proceedings international conference offshore site investigation and geotechnics, SUT 2002, London, Nov, pp 341–356

Remai Z (2013) Correlation of undrained shear strength and CPT resistance. Zsolt Rémai Periodica Polytechnica Civil Engineering 57(1):39–44

Rix GJ, Stokes KH (1992) Correlation of initial tangent modulus and cone resistance, Proceedings international symposium on calibration chamber testing, Postdam, NY, 1991, pp 351–362, Elsevier

Robertson PK (1990) Soil classification using the cone penetration test. Can Geotech J 27 (1):151–158

Robertson PK (2009) Interpretation of cone penetration tests – a unified approach. Can Geotech J 46:1337–1355

Robertson PK (2012) Interpretation of in-situ tests – some insights. In: J.K. Mitchell Lecture, Proceedings of ISC'4, Recife, Brazil

Robertson PK, Cabal KL (2010) Estimating soil unit weight from CPT. In: 2nd international symposium on cone penetration testing, CPT'10, Hungtington Beach, California

Robertson PK, Cabal KL (2014) Guide to cone penetration testing for geotechnical engineering, 6th edn. Gregg Drilling & Testing, Inc., Signal Hill, California

Robertson PK, Campanella RG (1983) Interpretation of cone penetration tests – Part I (sand). Can Geotech J 20(4):718–733

Robertson PK, Campanella RG (1995) Liquefaction potential of sands using the cone penetration test. J Geotech Eng ASCE 22(3):298–307

Robertson PK, Wride CE (1997) Cyclic liquefaction and its evaluation based on SPT and CPT. In: Proceedings of NCEER workshop on evaluation of liquefaction resistance of soils, National Center for Earthquake Engineering Research, State University of New York, Buffalo, Technical Report No. NCEER-97-0022, pp 41–88

Robertson PK, Wride CE (1998) Evaluating cyclic liquefaction potential using the cone penetrometer test. Can Geotech J 35(3):442–459

Robertson PK, Campanella RG, Wightman A (1983) SPT-CPT correlations. J Geotech Eng ASCE 109(11):1449–1459

Robertson PK, Campanella RG, Gillespie D, Greig J (1986) Use of piezometer cone data. In: Proceedings ASCE specialty conference in-situ'86 use of insitu testing in geotechnical engineering, pp 1263–1280

Sandevan R (1990) Strength and deformation properties of fine grained soils obtained from piezocone tests, PhD thesis, NTH Trondheim

Sanglerat G (1972) The penetrometer and soil exploration. Elsevier Pub, Amsterdam, 488 pp

Schmertmann JH (1970) Static cone to compute static settlement over sand. Journal of the Soil Mechanics and Foundations Division, ASCE, 96(SM3):1011–1043

Schnaid F (2009) In situ testing in goemechanics. Taylor and Francis, London, 329p

Seed HB, Idriss IM (1971) Simplified procedure for evaluating soil liquefaction potential. J Soil Mech Found Eng ASCE 97(SM9):1249–1273

Senneset K, Sandven R, Janbu N (1989) The evaluation of soil parameters from piezocone tests. In: Proceedings in situ testing of soil properties for transportation facilities, Research Council, Trans. Research Board, Washington DC

Sivakugan N, Das BM (2010) Geotechnical engineering, a practical problem solving approach. J Ross Publishing, Fort Lauderdale

Suzuki Y, Koyamada K, Tokimtsu K (1997) Prediction of liquefaction resistance based on CPT tip resistance and sleeve friction. In: Proceedings 14th international conference on soil mechanics and foundation engineering, Hamburg, Germany, vol 1, pp 603–606

Sykora DW, Stokoe KH II (1983) Correlations on in situ measurements in sands of shear wave velocity, soil characteristics, and site conditions, Geotechnical Engineering Report GR83-33. University of Texas at Austin, Austin

Villet WCB, Mitchell JK (1981) Cone resistance, relative density and friction angle, Cone Penetration Testing and Experience. Session at the ASCE National Convention, St Louis, pp 178–207, ASCE

Wair BR, DeJong JT, Shantz T (2012) Guidelines for estimation of shear wave velocity profiles, PEER report 2012/08. University of California, Pacific Earthquake Research Center, Berkeley

Youd TL, Idriss IM, Andrus RD, Arango I, Castro G, Christian JT, Dobry R, Finn WDL, Harder LF, Hynes ME, Ishihara K, Koester J, Liao S, Marcuson WF III, Martin GR, Mitchell JK, Moriwaki Y, Power MS, Robertson PK, Seed R, Stokoe KH (2001) Liquefaction resistance of soils: summary report from the 1996 NCEER and 1998 NCEER/NSF workshop on evaluation of liquefaction resistance of soils. ASCE J Geotechnical & Geoenvironmental Engineering 127:817–833

Chapter 6
Pressuremeter Test

Abstract This chapter focusses on another insitu test, the pressuremeter. The Menard type pressuremeter and the self boring pressuremeter are described including the test procedures and stress relaxation at the commencement of the test. The pressuremeter test has a strong theoretical base and this is described in the book. A description of obtaining the insitu lateral stress, modulus, undrained shear strength for cohesive soils and the friction angle for granular soils from theory is provided. Empirical correlations to correct the parameters obtained from a direct theoretical interpretation of the pressuremeter test curve are provided. The use of Menard type pressuremeter test results directly in the design of shallow foundations using empirical correlations is described. Correlations to obtain the ultimate bearing capacity and ultimate skin friction of deep foundations are also presented. Finally, correlations between the pressuremeter test results and SPT as well as the cone penetrometer test results are provided.

Keywords Pressuremeter • Correlations • In situ lateral stress • Modulus • Undrained shear strength • Bearing capacity

6.1 Pressuremeter Test – General

In situ devices are generally used in practice to obtain either the strength characteristics, deformation properties, or assess the in situ stress state of a soil deposit. The pressuremeter could be considered a unique device amongst the in situ devices because it has the potential to derive the full stress-strain curve. The test is also unique considering the range of materials it could be used compared to a Standard Penetration Test which cannot be used in rock, the vane shear test which can only be used in soft to firm clay, and the cone penetrometer test which also cannot be used in rock.

The original pressuremeter introduced to the world by Louis Menard in the early fifties was a simple device, which was easily operated. Its initial simplicity and convenience were not restricted to its construction and operation but to interpretation as well. The apparent versatility, simplicity and convenience created much interest among engineers and caught the attention of the researchers. Both these groups have been responsible for extensive modifications over the years, to bring this instrument to its present state and capable of being used in different deposits under variable conditions.

Although the concept of expanding a balloon like device within a borehole to derive deformation characteristics of soil is attributed to Kögler in 1933 (see Baguelin et al. 1978) the momentum for its development to be an alternative in situ test is attributed to Menard. Although the shape, size and varieties have been in the market since its introduction, the basic operational procedure has not changed significantly.

While the test itself is relatively simple there are operational difficulties and therefore operator skills play an important role compared to carrying out a vane shear test or other in situ test. The test results could be interpreted easily because the theory behind the test is very simple.

Only the basic principles of the test and the major modifications that have been developed since the fifties are presented here. If a more detailed description is required the reader is referred to Baguelin et al. (1978), Briaud (1992), Clarke (1995), Mair and Wood (1987) and Schnaid (2009).

6.1.1 Menard Type Pressuremeter

The standard pressuremeter is generally known as the Menard Pressuremeter and comprises two main components, viz., the control unit and the probe as shown in the schematic diagram in Fig. 6.1. The control unit consists of a gas supply and a device to control and measure volume changes in the probe. The probe is lowered into an existing borehole predrilled to a slightly larger diameter than the probe to the nominated depth of testing. The probe consists of three cylindrical chambers made out of rubber, physically adjoining each other, but operating separately. These cells are fixed to a hollow rod, which runs the length of the probe and keeps the cells in line. Only the middle chamber is used for measuring purposes and appropriately called the measuring cell. The two end chambers, known as guard cells, have been incorporated to reduce end effects on the measuring cell. The measuring cell is pressurized by water entering through the probe through leads from the control unit. The guard cells are pressurized by gas to the same pressure as the measuring cell, to create uniform pressure conditions along the total length of the probe. In addition, the guard cells restrain the axial expansion of the measuring cell into the borehole above and below the instrument. This eliminates a possible source of error in the volume measurement due to non-radial expansion of the membrane.

The standard pressuremeter test is run as a stress controlled test. Equal increments of pressure are applied to the probe, generally at one minute intervals. For a particular pressure step, the pressure and the volume of water injected to the measuring cell are determined at intervals of 30 s and 1 min after the pressure increment is applied. About 8–15 pressure increments are generally applied in a single test. After making corrections to the applied pressure and volume expansion

6.1 Pressuremeter Test – General

Fig. 6.1 Menard pressuremeter

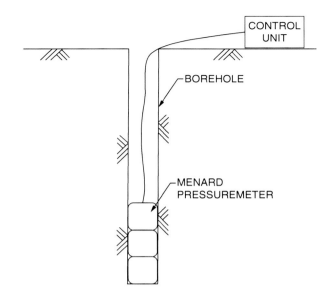

to account for the hydrostatic head in the measuring cell, the resistance and the membrane and expansion of the lead tubing, the resulting pressure-volume curve (i.e. one minute readings) will be as shown in Curve 1 in Fig. 6.2. It should be noted that conventionally, the pressure-volume response of a Menard pressuremeter is plotted with the pressure axis horizontal. Curve 2 in Fig. 6.2 shows the creep curve, obtained by plotting the change of volume between 30 s and 1 min readings for each pressure step.

There are three characteristics phases that can be identified:

Phase 1: The initial part of the curve (*OA*) represents the recompression of the soil. At point '*A*', the at-rest conditions are said to be re-established and are defined as V_0 and p_o which are taken as the initial volume of the cavity and the initial pressure on the borehole wall respectively.

Phase 2: The second phase shows a pseudo-elastic relationship between the pressure and the volume, and is indicated by *AB* in curve 1 in Fig. 6.2. This phase extends from (p_o, V_0) to (p_F, V_F) where '*F*' indicates the stage at which plasticity is initiated at the cavity wall. The value of p_F is generally known as the creep pressure. The point '*B*' is difficult to locate, in which case, the creep curve (Curve 2) is used to locate the creep pressure by defining p_F as the pressure at which an increase in creep volume due to plasticity begins.

Phase 3: The third phase illustrates the elasto-plastic nature of the soil medium. At this stage, an annulus of soil around the borehole is in a failed state while beyond that a non-failed region exists. The value p_L denotes the pressure at which infinite expansion occurs and p_l is defined as the pressure at which the initial volume doubles.

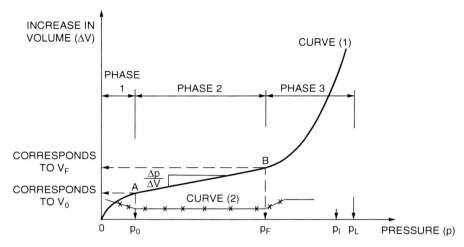

Fig. 6.2 Menard type pressuremeter curve

6.1.2 Self-Boring Pressuremeter

The standard Menard pressuremeter test is conducted in a pre-drilled borehole. The process has several issues because the soil being tested has been disturbed from its initial state, the disturbance originating from two sources:

- Disturbance at the wall due to drilling methods adopted including drill bits
- Disturbance due to soil relaxation

Both these were felt to have considerable influence on the derived parameters (Hartman 1974; Roy et al. 1975). The first of these could be minimized by the proper choice of drilling techniques and of drills to suit soil conditions. The second cause of disturbance, and probably the more important when careful drilling is employed, concerns the relaxation of soil stresses which occur in both cohesive and granular soils, once a borehole is created. The magnitude of disturbance can be illustrated by a simple example of considering an ideal soil medium having isotropic and homogeneous characteristics and undergoing an undrained pressuremeter test under plane strain conditions. Figure 6.3a shows the pressure-expansion curve, point 'O' representing in situ conditions. When a borehole is created the pressure at the borehole wall is reduced from in situ lateral stress (p_o) to zero (Line OC). Line OC would be linear if stress relief is linearly elastic and, if reloaded, as the soil elements are not yielded the stress path will retrace the path CO and continue in the direction OA. The corresponding stress-strain curve of a soil

6.1 Pressuremeter Test – General

Fig. 6.3 Effect of Borehole stress relief (**a**) Pressure – Expansion curve (**b**) Stress – Stain curve

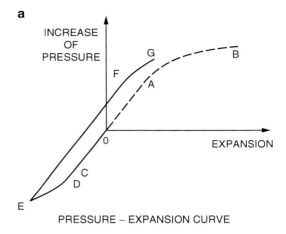
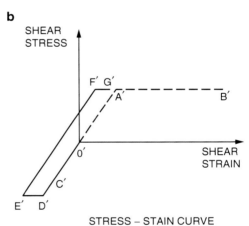

element at the borehole wall, for this unloading and unlading, is shown in Fig. 6.3b where the major points are denoted with a prime (') added to the letters used at the corresponding points in Fig. 6.3a. It is evident from both figures that if the soils behave in a linear elastic manner, on stress relief and subsequent reloading, at rest conditions are re-established. Therefore no apparent difference to the pressuremeter response would be observed on further loading.

If the soil medium had lower strength parameters compared to the in situ stresses (i.e., small values of $c_u/p_{o'}$), when stress relief occurs the soil may yield on reaching point 'D' in Fig. 6.3a (point 'D'' in Fig. 6.3b). If the soil behaves in an elastic-ideally plastic manner under Tresca criterion, the stress strain curve should proceed to point E' at which point the radial stress at the borehole wall is now zero (but not the circumferential stress). On reloading, the stress strain curve would be initially linear following $E'F'G'$ (corresponding to $ODEFG$ in Fig. 6.3a). The curve ODE

represents the unloading where DE is nonlinear because of failed elements in the surrounding soil, while EFG represents the reloading curve. Therefore it is evident that stress-relief would be non-linear because of plastic deformation with hysteresis. This produces a pressure-expansion curve quite different to an ideal curve with no stress-relief (i.e., curve OAB). This could be even worse if remoulding of soils occur. Therefore the derived parameters from the two curves cannot be expected to be similar.

Concerns of the influence of stress relief on measured properties led to research conducted to eliminate this undesirable effect. The result produced another significant development in pressuremeter technology with the emergence of the self-boring pressuremeter (Baguelin et al. 1972; Wroth and Hughes 1973). The self-boring pressuremeter was able to significantly reduce the disturbance created due to stress relief in a borehole.

Two groups are responsible for the development of the self-boring pressuremeter (SBP), one group based in France (Baguelin et al. 1978) and the other group in Cambridge (Wroth and Hughes 1973). The fundamental difference between the SBP and the conventional Menard type pressuremeter lies in the method of insertion. While the Menard pressuremeter is inserted into a predrilled borehole, the SBP, as the name implies, bores its own hole and thereby reduces the disturbance due stress relaxation in a pre-drilled borehole. The two groups, almost simultaneously produced the first workable self-boring pressuremeter. The French version of the self-borer is known as PAF (Baguelin et al. 1974) while the English version is called the Camkometer (Wroth and Hughes 1973). The main difference between the two types lies in the method of radial strain measurement; PAF measuring the volumetric changes and converting to a radial strain while the Camkometer using strain gauges to measure the radial strain at the wall of the borehole. For purposes of illustration only the Camkometer is described here.

The Camkometer consists of a single cell (i.e. no guard cells) and the bottom portion has a cutting edge which is beveled inside to make the borehole wall smooth on cutting. The self-boring mechanism consists of a cutter or a grinder rotating about the drill rod inside the cutting edge and the soil entering the cutting head is cut into pieces, and water or drilling fluid washes these cuttings up to the surface in slurry form. During insertion, the rubber membrane on the outside of the instrument is restrained to be the same diameter as the cutting head and once the instrument is at the nominated depth, the inflatable membrane is expanded against the borehole wall.

6.1.3 Other Developments

While the above describes the basic Menard type and self-boring pressuremeter, there have been many advances to the instruments. They include the push-in type pressuremeters, changes to measuring systems, pore pressure measurements, drilling procedures, sheathing of membranes to reduce damage to rubber membrane etc. The reader is directed to several books including Baguelin et al. (1972); Briaud (1992), Clarke (1995), Mair and Wood (1987) and Schnaid (2009).

Although the design, construction and operation of the pressuremeter have all been modified over the years, the basic principle has remained the same. The principle of expanding a cavity in a medium to determine its deformation and strength properties has an inherent appeal and therefore continues to create much promise with regard to geotechnical investigations.

6.2 Pressuremeter Test – Theoretical Interpretation

As previously discussed, the possibility of deriving the complete stress-strain curve of a soil from a pressuremeter test is a major advantage. However, its use for deriving relevant engineering properties is clearly a function of the relevancy of any analysis used to interpret the results obtained.

When a pressuremeter test is conducted in a medium of soil, the pressure-expansion curve produced can be analysed to derive many properties of the soil. The most widely derived parameters are:

1. in-situ lateral stress or K_o
2. modulus
3. undrained shear strength for cohesive soils
4. friction angle of granular soils

In addition to derivation of the above parameters based generally on cavity expansion theory, there have been correlations developed to directly assess foundation design parameters such as the ultimate bearing capacity of a shallow foundation or a deep foundation. It is found that empirical factors are used in most design procedures based on experience and calibrated laboratory and field tests. The empirical factors vary for the different types of pressuremeter as one would expect because the behavior is different.

As previously discussed, the Menard type pressuremeter (MPM) and the self boring pressuremeter (SBP) are the widely used pressuremeter types in the industry and our attention is limited to these two types.

6.3 Parameter Derivation

6.3.1 In-Situ Lateral Stress

Although the assessment of in situ vertical stress could be easily assessed, the evaluation of the horizontal stress is more difficult because of its heavy dependence on the nature of the deposit, its origin and changes that have occurred over the years. The in situ lateral stress (or the coefficient of earth pressure at rest, K_o) of a soil medium is one of the most important parameters in the field of geotechnical engineering.

There are several methods of obtaining the in situ later stress or K_o of a soil deposit and can be broadly classified as follows:

1. Indirect methods
2. Direct methods

Indirect methods include semi-empirical or empirical relationships (Jaky 1944; Brooker and Ireland 1965) and laboratory tests such as consolidation or triaxial tests. As Wroth (1975) pointed out, most of the investigations carried out to obtain semi-empirical or empirical relationships have been based on laboratory remoulded samples and therefore fail to account for naturally occurring phenomena such as the nature of deposition. Thus any parameters derived will not be truly representative of in-situ conditions. However they are useful in understanding of the soil conditions and for identifying lower and upper bounds (Tavenas et al. 1975). Although laboratory tests on "undisturbed" samples could be used to predict K_o, the major disadvantage is the disturbance that obviously creeps into "undisturbed" samples at sampling stage, stress relaxation and manual handling in the laboratory (Poulos and Davis 1972).

The above concerns on indirect methods have created more attention to direct methods for the measurement of K_o. The pressuremeter is one of the main in situ tests that could be used to obtain K_o. It is capable to handle soft to stiff clays, loose to dense sands, and soft rocks. This statement is probably correct for the self-boring pressuremeter because the stress relief is not significant if the test is carried out correctly (see typical curve in Fig. 6.3). However, because the Menard type pressuremeter is used in a pre-drilled borehole, the in-situ conditions could be greatly affected by the drilling process, and the disturbance and stress relief at the borehole except perhaps for very stiff to hard materials including rock. The disturbance due to stress relief could be expected to be higher in softer soils because they have lower shear strength parameters and therefore would reach failure at lower values of stresses. Also, softer soils are liable to remould on stress relief. Tavenas et al. (1975) report unacceptable values of K_o predicted by the Menard type pressuremeter for softer soils.

The most convenient method to interpret the in situ lateral stress from the pressuremeter test is to assume that it is equal to the pressure at the beginning of the linear portion of the test curve (p_0 in Fig. 6.2). However, as Mair and Wood (1987) point out this is incorrect and should not be adopted. Several researchers have developed methods to better predict the in situ horizontal pressure. Readers are referred to Clarke (1995) who discusses several methods including the method proposed by Marsland and Randolph (1977).

On the other hand, the self-boring pressuremeter is one of the most suited in situ tests to obtain the in situ lateral stress (p_0) or Ko. This is because the disturbance it creates on insertion is significantly less than some of the other in situ tests including the Menard type pressuremeter. According to Mair and Wood (1987), deflection at lift off by visual inspection of the pressuremeter curve is the most reliable method to assess p_o in clays.

6.3.2 Young's Modulus

The Young's modulus is derived from the initial part of the pressuremeter curve where the material is assumed to behave in a linear elastic manner. It can be shown (Gibson and Anderson 1961) that the following equation could be derived from cavity expansion theory for small strains:

$$G = \frac{\Delta p}{\Delta V/V} \qquad (6.1)$$

where V is the current volume and ΔV the increase in volume (see Fig. 6.2) and G is the shear modulus. For small strains, the above equation could be re-written as follows:

$$G = \frac{\Delta p}{\Delta V/V_o} \qquad (6.2)$$

where V_0 is the initial cavity volume.

Therefore from the gradient from the pressuremeter curve (p Versus ΔV) in Fig. 6.2, the shear modulus could be obtained as V_0 is known. This equation is applicable to both drained and undrained conditions.

The above procedure works well for a self-boring pressuremeter whereas significant underestimation is expected because of the stress relief involved in a Menard type pressuremeter test. Therefore, some practitioners use the reload modulus. Mair and Wood (1987) state that the unload-reload modulus would give a more reasonable estimation of the elastic properties than the initial modulus from a Menard pressuremeter curve. However, Briaud in his Menard lecture (2013) pointed out that the reload modulus depends on the strain amplitude over which the unload/reload loop is performed and unless the strain amplitude and stress level adopted during the test matches the application different results could be obtained. Briaud (2013) performs a loop at the end of the linear phase and unload until the pressure is reduced to 50 % of the peak pressure but strongly discourages the use of the reload modulus.

Menard and his colleagues identified the issue very early and in France an empirical approach has been used in practice. Assuming $\nu = 0.33$, Eq. (6.1) is converted to the following:

$$E_M = 2.66 V_m \Delta p/\Delta V \qquad (6.3)$$

Table 6.1 Empirical factor α for various soils

	Peat		Clay		Silt		Sand		Sand and gravel	
	E_M/p_l	α	E_M/p_l	α	E_M/p_l	α	E_M/p_l	α	E_M/p_l	α
OC			>16	1	>14	2/3	>12	1/2	>10	1/3
NC		1	9–16	2/3	8–14	1/2	7–12	1/3	6–10	1/4
Weathered and remoulded			7–9	1/2		1/2		1/3		1/4

After Baguelin et al. (1978)
*p_l is the limit pressure at which the initial volume doubles – see Fig. (6.2)

where E_M is known as Menard modulus and V_m refers to the volume at the mid point between V_0 and V_F in Fig. 6.2. They advocated that Menard modulus should be increased by the use of an empirical factor α to obtain the Young's modulus, E.

$$E = E_m^{1/\alpha} \tag{6.4}$$

This factor α is a function of material properties as given in Table 6.1 for soils.

6.3.3 Undrained Shear Strength in Clay

Baguelin-Ladanyi-Palmer (B-L-P) theory was developed simultaneously, but independently by Baguelin, Jézequel, Lemée and Le Méhauté group, Ladanyi and Palmer, all in 1972. The B-L-P theory is applicable to undrained expansion in saturated cohesive soils and could be used to derive the complete stress-strain curve. Although the usual assumptions of an isotropic, homogeneous medium and plane strain expansion are applied, the only assumption regarding the stress strain model is that all elements of soil will behave according to a unique stress-strain model. The derivation of equations is described in the technical papers by Baguelin et al. (1972), Ladanyi (1972) and Palmer (1972).

For the nonlinear part of the pressure expansion curve the following equation is derived.

$$\frac{\sigma_r - \sigma_\theta}{2} = \varepsilon_o(1+\varepsilon_o)(1+\varepsilon_o/2)\frac{dp}{d\varepsilon_o} \tag{6.5}$$

where

$\sigma_r - \sigma_\theta/2$ = shear stress; σ_r the radial stress and σ_θ the circumferential stress
ε_o = cavity strain at the borehole wall
p = applied pressure at the borehole wall

Equation (6.5) reduces to the following for small strains:

6.3 Parameter Derivation

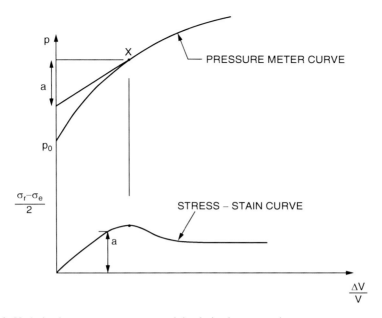

Fig. 6.4 Undrained pressuremeter curve and the derived stress-strain curve

$$\frac{\sigma_r - \sigma_\theta}{2} = \varepsilon_o \frac{dp}{d\varepsilon_o} \quad (6.6)$$

Equation (6.5) could also be written using volumetric strain as shown in Eq. (6.7).

$$\frac{\sigma_r - \sigma_\theta}{2} = \left(\frac{\Delta V}{V}\right) \frac{dp}{d\left(\frac{\Delta V}{V}\right)} \quad (6.7)$$

Equations (6.6) and (6.7) describe the shear strength of an element at the borehole wall during the expansion of the pressuremeter and could be obtained by plotting gradients to the pressuremeter curve. If for instance Eq. (6.7) is used, referring to Fig. 6.4, by drawing a tangent at a particular strain, X, the abscissa it makes on the pressure axis ('a') is equivalent to the right hand side of Eq. (6.6). Therefore by plotting 'a' directly beneath the point X, the stress-strain curve could be obtained.

The other method of obtaining the shear strength is the use of limit pressure, p_L by extrapolating the pressuremeter curve. As Mair and Wood (1987) point out, strengths obtained from limit pressure method appear to be less sensitive to the assumed reference conditions, and hence less sensitive to disturbance at the borehole wall. Mair and Wood (1987) also point out that the strengths obtained from a pressuremeter test is appreciably higher than those obtained from vane shear or laboratory tests. Schnaid (2009) iterates this observation when predicting undrained strengths from self-boring pressuremeter tests. He attributes this partially to end effects of the probe.

In Menard type pressuremeter tests generally an empirical approach is adopted which has some theoretical basis. If an elastic perfectly plastic soil is assumed, the following equation could be derived if undrained conditions (i.e. volume change is zero) are assumed (Gibson and Anderson 1961):

$$p = p_o + c_u + c_u \log_e \left\{ \left(\frac{G}{c_u}\right) \Delta V/V - \left(1 - \Delta V/V\right)\frac{p_o}{c_u} \right\} \quad (6.8)$$

where G/c_u is generally known as the rigidity index. As infinite expansion occurs when $\Delta V/V = 1$, the equation can be re-written as follows:

$$p_L = p_o + c_u + c_u \log_e \left(\frac{G}{c_u}\right) \quad (6.9)$$

The above equation could be re-written as follows:

$$p_L - p_o = N_c c_u \quad (6.10)$$

which is similar in formation to the bearing capacity or cone penetrometer equations, where

$$N_c = 1 + \log_e \left(G/c_u\right) \quad (6.11)$$

The factor N_c can be expressed in terms of Young's modulus (E) as follows if the Poisson's ratio is assumed to be 0.5 for undrained conditions:

$$N_c = 1 + \log_e \left(E/3c_u\right) \quad (6.12)$$

As previously mentioned, since an infinite expansion cannot be obtained in a field pressuremeter test, p_L cannot be directly measured. Therefore, it is the standard practice to modify Eq. (6.10) by the introduction of p_l, the pressure at which the initial volume doubles, i.e. $\Delta V/V_0 = 1$ (see Fig. 6.2). Then Eq. (6.10) could be written as follows by adopting a modified factor.

$$p_l - p_o = N_c^* c_u \quad (6.13)$$

As Baguelin et al. (1978) point out the rigidity index could have a wide range for a clay, between 200 and 2000, and therefore the modified factor N_c^* could have a wide range. They assessed the results of published and unpublished literature and found the modified factor to vary between 6.5 and 12 in the stiff to very stiff strength range with an average value of 9.

6.3 Parameter Derivation

Table 6.2 Empirical relationships between p_l and c_u

c_u	Clay type	References
$(p_l-p_o)/k$	$k = 2$ to 5	Menard (1957)
$(p_l-p_o)/5.5$	Soft to firm clays	Cassan (1972) and Amar and Jézéquel (1972)
$(p_l-p_o)/8$	Firm to stiff clays	
$(p_l-p_o)/15$	Stiff to very stiff clays	
$(p_l-p_o)/6.8$	Stiff clays	Marsland and Randolph (1977)
$(p_l-p_o)/5.1$	All clays	Lukas and LeClerc de Bussy (1976)
$(p_l-p_o)/10+25$		Amar and Jézéquel (1972)
$(p_l-p_o)/10$	Stiff clays	Martin and Drahos (1986)
$p_l/10+25$	Soft and stiff clay	Johnson (1986)

After Clarke (1995)

Empirical relationships have been proposed by several other groups and Clarke (1995) provides a summary which is shown in Table 6.2.

According to Baguelin et al. (1978), the Centre d'Etudes Me'nard proposed a value of 5.5 for the modified rigidity index in calculating the ultimate or residual strength ($c_{u\,res}$) of a clay:

$$p_l - p_o = 5.5 c_{u\,res} \qquad (6.14)$$

6.3.4 Friction Angle in Sands

In 1977, Hughes et al. proposed a method to obtain the friction angle of sands under drained conditions for the self-boring pressuremeter. Their tests on sands in the laboratory agreed well with the stress dilatancy theory of Rowe (1962). Using this theory and assuming that the angle of friction at failure is equal to the angle of friction at constant volume ($\phi'_{cv'}$), Eq. (6.15) was derived. Clarke and Gambin (1998) state that it is applicable only to dense sands although a correction could be applied from tests conducted in loose sands (Robertson and Hughes 1986).

$$\ln\left(p'\right) = s\ln\varepsilon_o + A \qquad (6.15)$$

The expression is a straight line in logarithmic scale and the gradient s is defined by the following:

$$s = \frac{(1+\sin\psi)\sin\phi'}{1+\sin\phi'} \qquad (6.16)$$

where ϕ' = friction angle and ψ' = angle of dilation.

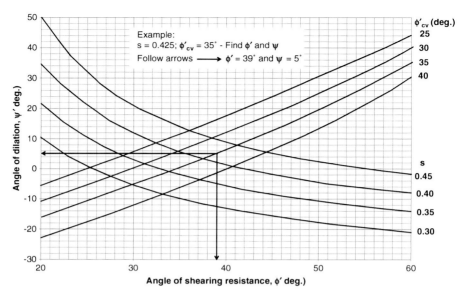

Fig. 6.5 Derivation of Friction angle and angle of dilation (Adapted from Mair and Wood 1987)

To obtain ϕ' and ψ', the following stress dilatancy relationship of Rowe (1962) is used:

$$\frac{1 - \sin\phi'}{1 + \sin\phi'} = \left(\frac{1 - \sin\phi'_{cv}}{1 + \sin\phi'_{cv}}\right)\left(\frac{1 - \sin\psi'}{1 + \sin\psi'}\right) \quad (6.17)$$

If ϕ'_{cv} can be found in the laboratory by testing a disturbed sample, from Eqs. (6.16) and (6.17), both ϕ' and ψ' could be found by solving the equations or using a chart presented by Mair and Wood (1987) and shown in Fig. 6.5. Mair and Wood (1987) comment that in the absence of test data, approximate values could be taken (based on experience) and the range $30° < \phi' < 35°$ covers most quartz sands and the uncertainty of $5°$ in ϕ'_{cv} corresponds to an uncertainty of about $2.5°$ in ϕ'.

The above relates to tests carried out by a self-boring pressuremeter. In a Menard type pressuremeter, as the disturbance is significant, the use of the above theory to derive ϕ' and ψ' is not recommended. An empirical method of determining the friction angle of a granular soil has been adopted by Menard and his colleagues in France which is based on the following equation (Baguelin et al. 1978).

$$p_l^{'*} = b\, 2^{(\phi' - 24)/4} \quad (6.18)$$

where

$p_l^{'*}$ = net effective limit pressure at which the initial volume of the probe doubles
b = constant = 2.5 on average

= 1.8 for wet soil with loose structure
= 3.5 for dry, structured soil

However, the accuracy of the empirical method does not appear to provide a high confidence in the derivation of friction angle in sands.

6.4 Correlations with Other Tests

6.4.1 Correlation Between Limit Pressure from Menard Type Pressuremeter and q_c from Cone Penetrometer Test

Amar et al. (1991) reports the following relationship given in Table 6.3 (After Van Wambeke and d'Hemricourt J 1982).

6.4.2 Correlations with Other Soil Parameters – Menard Type Pressuremeter

Briaud (2013) provided several correlations (see Tables 6.4 and 6.5) for the Menard type pressuremeter based on a data base of 426 tests carried out at 36 sites in sand, 44 sites in clay and 2 silt sites, and reported in Briaud et al. (1985) and Briaud (1992). The latter comments on the significant scatter and cautions on their use but admits they would be very useful in preliminary calculations and for estimate purposes.

6.5 Use of Menard Type Pressuremeter Test Results Directly in Design

The Menard Pressuremeter (MPM) has been extensively used in France in site investigations for various types of foundations and semi empirical design rules have been developed, based partly on theory and partly on observations of foundation behaviour. Menard, the inventor of the MPM, commenced the establishment of design rules so that what is measured could be directly related to design of foundations. After significant research carried out in France the latest design rules for the MPM test were incorporated in a Code of Practice called "Fascicule 62 – Titre V", was approved and officially adopted by MELT

Table 6.3 Correlation between limit pressure and q_c

Soil type	Clay	Silt	Sand	Dense sand and gravel
q_c/p_l	3	6	9	12

After Van Wambeke and d'Hemricourt J (1982)

Table 6.4 Correlations for Sand (Column A = Number in Table × Row B)

Column A = Number in Table × Row B

A \ B	E_o (kPa)	E_R (kPa)	p_l^* (kPa)	q_c (kPa)	f_s (kPa)	SPT N
E_o (kPa)	1	0.125	8	1.15	57.5	383
E_R (kPa)	8	1	64	6.25	312.5	2174
p_l^* (kPa)	0.125	0.0156	1	0.11	5.5	47.9
q_c (kPa)	0.87	0.16	9	1	50	436
f_s (kPa)	0.0174	0.0032	0.182	0.02	1	9.58
SPT N	0.0026	0.00046	0.021	0.0021	0.104	1

After Briaud (2013)

Table 6.5 Correlations for Clay (Column A = Number in Table × Row B)

Column A = Number in Table × Row B

A \ B	E_o (kPa)	E_R (kPa)	p_l^* (kPa)	q_c (kPa)	f_s (kPa)	c_u (kPa)	SPT N
E_o (kPa)	1	0.278	14	2.5	56	100	667
E_R (kPa)	3.6	1	50	13	260	300	2000
p_l^* (kPa)	0.071	0.02	1	0.2	4	7.5	50
q_c (kPa)	0.40	0.077	5	1	20	27	180
f_s (kPa)	0.079	0.0038	0.25	0.05	1	1.6	10.7
c_u (kPa)	0.010	0.0033	0.133	0.037	0.625	1	6.7
SPT N	0.0015	0.0005	0.02	0.0056	0.091	0.14	1

After Briaud (2013)

(ministere de l'Equipment, du Logement et des Transports) in March 1993 (Frank 2009). In the sub-sections below a summary of most used parameters are discussed based on Frank (2009).

6.5.1 Ultimate Bearing Capacity (q_u) of Shallow Foundations – Menard Type Pressuremeter

The ultimate bearing capacity is related to the ultimate limit pressure p_l (i.e. pressure at which the initial volume doubles) and the ultimate bearing capacity equation for a vertically loaded foundation is written in the simple form:

6.5 Use of Menard Type Pressuremeter Test Results Directly in Design

Table 6.6 MPM Bearing factor for the design of shallow foundations

Soil & Category[a]	k_p
Clay & Silt A, Chalk A	$0.8\{1 + 0.25(0.6 + 0.4^{b/L})\}^{D_e/b}$
Clay & Silt B	$0.8\{1 + 0.35(0.6 + 0.4^{b/L})\}^{D_e/b}$
Clay C	$0.8\{1 + 0.50(0.6 + 0.4^{b/L})\}^{D_e/b}$
Sand A	$\{1 + 0.35(0.6 + 0.4^{b/L})\}^{D_e/b}$
Sand & Gravel B	$\{1 + 0.50(0.6 + 0.4^{b/L})\}^{D_e/b}$
Sand & Gravel C	$\{1 + 0.80(0.6 + 0.4^{b/L})\}^{D_e/b}$
Chalk B & C	$1.3\{1 + 0.27(0.6 + 0.4^{b/L})\}^{D_e/b}$
Marl & Calcareous Marl & Weak Rock	$\{1 + 0.27(0.6 + 0.4^{b/L})\}^{D_e/b}$

[a]Category defined in Table 6.7

Table 6.7 Categories for soil and rock

Soil	Category	Consistency/density	p_l (MPa)
Clay & Silt	A	Soft	0.7
	B	Stiff	1.2–2
	C	Hard (clay)	>2.5
Sand & Gravel	A	Loose	<0.5
	B	Medium	1–2
	C	Dense	>2.5
Chalk	A	Soft	<0.7
	B	Weathered	1–2.5
	C	Dense	>3
Marl & Calcareous Marl	A	Soft	1.5–4
	B	Dense	>4.5
Weak Rock	A	Weathered	2.5–4
	B	Fragmented	>4.5

$$q_u - q_o = k_p(p_l - p_o) \tag{6.19}$$

where

k_p = Menard Type Pressuremeter bearing factor
q_o and p_o = confining vertical and horizontal stresses
$p_l - p_o$ = net limit pressure, p_l^*

The bearing factor k_p is a function of the type and consistency/density of the soil, relative embedment D_e/b (D_e is the equivalent embedment depth) and of the width/length (b/L) ratio. A summary is presented in Table 6.6 for shallow foundations and the categories are presented in Table 6.7. Schnaid (2009) reports that MELT approach requires the net limit pressure to be taken to be the equivalent value over a zone within $1.5\,B$ of the foundation level where B is the width of the footing.

Schnaid also discusses the MELT requirements for inclined loads and foundations adjacent to slopes.

6.5.2 Ultimate Bearing Capacity of Deep Foundations – Menard Type Pressuremeter Test

Equation (6.19) is still valid for the assessment of the ultimate bearing capacity of a deep foundation with the end bearing factor k_p shown in Table 6.8. Schnaid (2009) describes the Menard approach to calculate the equivalent net limit pressure of a pile having an equivalent diameter 'B' equal to $4A/P$, A being the cross sectional area and P the perimeter of the pile. The procedure involves the integrated average over a zone from the base of pile:

- extending upwards to 'b', and
- extending downward to '$3a$'

where

'a' is *equal to B/2 or* 0.5 m *whichever is greater*
'b' is the minimum of 'a' *and* 'h' *where* 'h' *is the embedment in the bearing layer*

6.5.3 Skin Friction for Deep Foundations – Menard Type Pressuremeter

Figure 6.6 provides the limit unit skin friction values for bored and driven piles and Table 6.9 provides the category based on the type of soil, type of pile and also construction conditions (see Frank 2009).

6.5.4 Correlation with q_c and SPT N

Bustamante and Gianeselli (1993) states that, in France, although SPT tests are considered totally unsuitable for some soils, they are not ignored because of he versality and because in many countries SPT is the only test available at preliminary stages. For practical reasons they present a correction chart for SPT N value and q_c from CPT against pressuremeter test results in Table 6.10.

6.5 Use of Menard Type Pressuremeter Test Results Directly in Design

Table 6.8 End bearing factors for deep foundations (k_p)

Soil	Category[a]	k_p Non displacement	displacement
Clay & Silt	A	1.1	1.4
	B	1.2	1.5
	C	1.3	1.6
Sand & Gravel	A	1.0	4.2
	B	1.1	3.7
	C	1.2	3.2
Chalk	A	1.1	1.6
	B	1.4	2.2
	C	1.8	2.6
Marl & Calcareous Marl	A	1.8	2.6
	B		
Weak Rock	A	1.1–1.8	1.8–3.2
	B		

[a]Category defined in Table 6.7

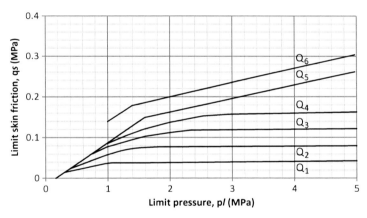

Fig. 6.6 Menard type pressuremeter – limit skin friction (Adapted from Bustamante and Gianeselli 1981)

6.5.5 Other Design Parameters from Menard Type Pressuremeter

Many other design procedures related to shallow foundations, deep foundations, retaining walls, stone columns etc are available. The reader is directed to Baguelin et al. (1978), Briaud (1992), Clarke (1995), Frank (2009) and Schnaid (2009).

Table 6.9 MPM skin friction – classification (to be read with Fig. 6.6)

Type of pile	Clay & Silt			Sand & Gravel			Chalk			Marl		Rock
	A	B	C	A	B	C	A	B	C	A	B	
Bored without casing	Q_1	Q_1 $Q_2(1)$	Q_2 $Q_3(1)$				Q_1	Q_3	Q_4 $Q_5(1)$	Q_3	Q_4 $Q_5(1)$	Q_6
Bored under slurry	Q_1	Q_1 $Q_2(1)$		Q_1	Q_1 $Q_1(2)$	Q_3 $Q_2(2)$	Q_1	Q_3	Q_4 $Q_5(1)$	Q_3	Q_4 $Q_5(1)$	Q_6
Bored with temporary casing	Q_1	Q_1		Q_1	Q_1	Q_3	Q_1	Q_2	Q_3	Q_3	Q_4	–
		$Q_2(3)$			$Q_1(2)$	$Q_2(2)$			$Q_4(3)$			
Bored with permanent casing	Q_1			Q_1		Q_2	(4)			Q_2	Q_3	–
Piers (5)	Q_1	Q_2	Q_3	–			Q_2	Q_3		Q_4	Q_5	Q_6
Close ended steel tube	Q_1	Q_2		Q_2		Q_3	(4)			Q_3	Q_4	Q_4
Driven precast concrete	Q_1	Q_2		Q_3			(4)			Q_3	Q_4	Q_4
Driven cast-in-place	Q_1	Q_2		Q_2		Q_3	Q_2	Q_3		Q_3	Q_4	–
Concrete driven steel (6)	Q_1	Q_2		Q_3		Q_4	(4)			Q_3	Q_4	–

Notes: (1) rebored and grooved borehole wall (2) for long piles (longer than 30 m) (3) dry boring, no rotation of casing (4) in chalk, below the water table, where q_s can be significantly affected, a specific preliminary study is needed; full-scale test is recommended (5) bored in dry soils or rocks above the water table without borehole support (6) a preformed steel pile of tubular or H-section, with an enlarged shoe, is driven with simultaneous pumping of concrete (or mortar) into the annular space

For bored and grouted piles, readers are referred to Frank (2009)

Table 6.10 Chart to convert p_l and/or q_c in MPa to SPT N value

p_l(MPa)	0	1	2	3	
Clay or clayey silt	0	3	6	9	q_c
	0	15	30	45	N
Sand or gravel	0	8	16	24	q_c
	0	20	40	60	N
Marls	0	3.5	7	10.5	q_c
	0	20	40	60	N
Chalk	0	4	8	12	q_c
	0	6-12	12-24	18-36	N

References

Amar S, Jézéquel FJ (1972) Essais en place et en Laboratoire sur sols coherents comparison des resultats, Bulletin de Liaison de LCPC, Paris, No 58, pp 97–108

Amar A, Clarke BGF, Gambin MP, Orr TL (1991) The application of pressuremeter test results to foundation design in Europe. A state-of-the-art report by the ISSMFE European Technical Committee on Pressuremeters

Baguelin F, Jézéquel JF, Le Mée E, Le Méhauté A (1972) Expansion of cylindrical probes in cohesive soils. J Soil Mech Found Eng Div ASCE 98(SM11):1129–1142

Baguelin F, Jézéquel JF, Le Méhauté A (1974) Self-boring placement method of soil characteristics measurements. In: Proceedings conference on subsurface exploration for underground excavation and heavy construction, ASCE, Henniker N.H., pp 312–332

Baguelin F, Jézéquél JF, Shileds DH (1978) The pressuremeter and foundation engineering. Trans Tech Publications, Clausthal

Briaud JL (1992) The pressuremeter. AA Balkema, Rotterdam

Briaud JL (2013) The pressuremeter test: expanding its use, Menard Lecture. In: Proceedings international conference on soil mechanics and geotechnical engineering, Paris, vol 1, pp 107–126

Briaud JL, Noubani A, Kilgore J, Tucker LM (1985) Correlation between pressuremeter data: other parameters. Research report, Civil engineering, Texas A&M University

Brooker EW, Ireland HO (1965) Earth pressure at rest related to stress history. Can Geotech J 2(1):1–15

Bustamante M, Gianeselli L (1981) Prevision de la capacite portante des pieux isoles sous charge verticale. Regles pressiometriques et penetrometriques. Bull. Liaison Labo. P. et Ch., No.113, Ref. 2536, 83–108 (in French)

Bustamante M, Gianeselli L (1993) Design of auger displacement piles from in situ tests. In: 2nd international geotechnical seminar on deep foundations on bored and auger piles, Ghent (Belgium), pp 21–34, Balkema

Cassan M (1972) Corrélation entre essais in situ en méchanique des sols. Internal report, Fondasol, Avignon

Clarke BG (1995) Pressuremeters in geotechnical design. Blackie Academic & Professional, Glasgow

Clarke BG, Gambin MP (1998) Pressuremeter testing in onshore ground investigations. A Report by the ISSMGE Committee TC 16. In: International conference on geotechnical site characterization, Atlanta

Frank R (2009) Design of foundations in France with the use of Menard pressuremeter tests (MPM). Soil Mech Found Eng 46(6):219–231

Gibson RE, Anderson WF (1961) In situ measurement of soil properties with the pressuremeter. Civil Eng Public Works Rev 56(658):615–618

Hartman JP (1974) Finite element parametric study of vertical influence factors and the pressuremeter test to estimate the settlement of footings in sand. PhD thesis, University of Florida.

Hughes JMO, Wroth CP, Windle D (1977) Pressuremeter tests in sands. Geotechnique 27(4):455–477

Jaky J (1944) The coefficient of earth pressure at rest. J Soc Hung Archit Eng 7:355–358

Johnson LD (1986) Correlation of soil parameters from in situ and laboratory tests for building. In: Proceedings ASCE specialty conference in situ '86: use of in situ tests in geotechnical engineering, Blacksburg Virginia, pp 635–648

Ladanyi B (1972) In-situ determination of undrained stress-strain behavior of sensitive clays with pressuremeter. Can Geotech J 9(3):313–319

Lukas GL, LeClerc de Bussy B (1976) Pressuremeter and laboratory test correlations for clays. J Geotech Eng Div ASCE 102(GT9):954–963

Mair RJ, Wood DM (1987) Pressuremeter testing – methods of and interpretation. Butterworth, London

Marsland A, Randolph MF (1977) Comparison of the results from pressuremeter tests and large in situ plate tests in London Clay. Geotechnique 27(2):217–243

Martin RE, Drahos EG (1986) Pressuremeter correlations for preconsolidated clay. In: Proceedings ASCE specialty conference in situ '86: use of in situ tests in geotechnical engineering, Blacksburg Virginia, pp 206–220

MELT-Minist re de l'Equipement, du logement et des transports (1993). R gles Techniques de Conception et de Calcul des Fondations des Ouvrages de Genie Civil (Technical rules for the design of foundations of civil engineering structures). Cahier des clauses techniques generales applicables aux marches publics de travaux, FASCICULE N°62 -Titre V, Textes Officiels N° 93–3 T.O., 182 p

Menard L (1957) Mesures in situ des propriétés physiques des sols. Annales des Ponts et Chaussées, Paris, No. 14, pp 357–377

Palmer AC (1972) Undrained plane-strain expansion of a cylindrical cavity in clay: a simple interpretation of the pressuremeter test. Geotechnique 22(3):451–457

Poulos HG, Davis EH (1972) Laboratory determination of in situ horizontal stress in soil masses. Geotechnique 22:177–182

Robertson PK, Hughes JMO (1986) Determination of properties of sand from self-boring pressuremeter tests. In: Proceedings 2nd international symposium on pressuremeter marine applications, Texam, USA, ASTM No STP 950, pp 283–302

Rowe PW (1962) The stress-dilatancy relation for static equilibrium of an assembly of particles in contact. Proc R Soc Lond A269:500–527

Roy M, Juneau R, La Rochelle P, Tavenas F (1975) In-situ measurement of the properties of sensitive clays by pressuremeter tests. In: Proceedings conference in situ measurement of soil properties, Raleigh, vol 1, pp 350–372

Schmertmann JH (1975) Measurement of in situ shear strength, Keynote lecture, In: Proceedings of the conference on in-situ measurement of soil properties, ASCE, vol 2, pp 57–138

Schnaid F (2009) In situ testing in geomechanics, the main tests. Taylor and Francis, London/New York

Tavenas FA, Blanchette G, Leroueil S, Roy M, LaRochelle P (1975) Difficulties in in-situ determination of K_o in soft sensitive clays. In: Proceedings conference on in situ measurement of soil properties, Raleigh, vol 1, pp 450–476

Van Wambeke A, d'Hemricourt J (1982) Correlation between the results of static and dynamic probings and pressuremeter tests. In: Proceedings 2nd ESOPT, Balkema, Rotterdam

Wroth CP (1975) In situ measurement of initial stresses and deformation characteristics. In: Proceedings conference on in situ measurement of soil properties, vol 1, pp 181–230

Wroth CP, Hughes JMO (1973) An instrument for the in-situ measurement of the properties of soft clays. In: Proceedings 8th international conference soil mechanics and foundation engineering, Moscow, vol 1.2, pp 487–494

Chapter 7
Dilatometer Test

Abstract Dilatometer is one of the latest arrivals among the in situ testing devices. It has become more versatile with the development of seismic dilatometer, which also measures the shear wave velocity. The three major indices computed in the dilatometer test are the material index I_D, horizontal stress index K_D, and dilatometer modulus E_D. These three are used in identifying the soil type and determining the soil parameters and this chapter provides correlations to obtain these parameters.

Keywords Dilatometer • Correlations • Coefficient of earth pressure at rest • Over consolidation ratio • Constrained modulus • Friction angle

7.1 Introduction

The flat dilatometer (ASTM 6635) was developed by Dr. Sylvano Marchetti in 1975 in Italy. It consists of a 240 mm long, 95 mm wide and 15 mm thick stainless steel blade with a flat, thin and expandable 60 mm diameter and 0.20–0.25 mm thick circular steel membrane that is mounted flush with one face (See Fig. 7.1). The blade has a cutting edge at the bottom end, tapered over 50 mm, with an apex angle of 24°–32°.

The blade is generally pushed into the ground by the penetration test rig, at a rate of 20 mm/s. Sometimes, impact driven hammers, similar to those used to drive the standard penetration test split-spoon sampler, are also being used. One of the advantages of DMT is the wide variety of equipment and techniques available for pushing the blade into the ground. Nowadays, it is suggested that the 20 mm/s penetration rate does not have to be maintained. When using a 20 tonne penetrometer truck, it is possible to achieve 100 m of profiling in a day. A gas pressure unit at the surface is used to inflate the steel membrane when it is pushed into the ground.

When the dilatometer blade is pushed into the ground, three pressure readings are taken in a sequence:

(a) The pressure required to bring the membrane in flush with the soil (i.e., to just move the membrane), known as lift-off pressure or "A pressure" (See Fig. 7.1a);

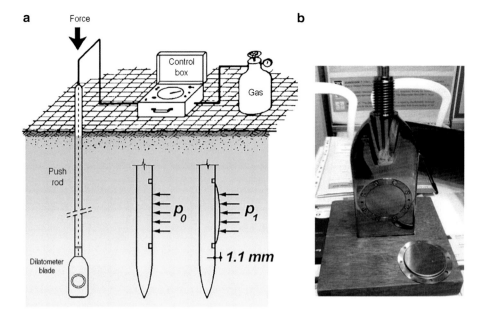

Fig. 7.1 (a) Schematic diagram of a dilatometer test (Adapted from Marchetti 2001) and (b) Photograph of a dilatometer blade and the steel membrane

(b) The pressure required to expand the membrane against the soil laterally by 1.1 mm, known as "B pressure"; and
(c) The pressure when the membrane is deflated, known as the closing pressure or "C pressure." This is an optional reading.

The first two are the most used in the computations. C pressure is used for determining the pore water pressure in the ground. Pressures A, B and C are corrected for the membrane stiffness, determined through calibration. These three corrected values are denoted by p_0, p_1 and p_2, respectively. The test is generally carried out at 200 mm depth intervals.

In stiffer soils, p_0 and p_1 can be determined as (Marchetti 1980)

$$p_0 = A + \Delta A \quad (7.1)$$

$$p_1 = B - \Delta B \quad (7.2)$$

where, ΔA and ΔB are the calibration corrections applied on the A and B pressures. ΔA is the external pressure required on the membrane in free air to collapse it against its seating, in overcoming the membrane stiffness. It is determined through applying suction to the membrane. ΔB is the internal pressure which in free air would lift the centre of the membrane by 1.1 mm from its seating, thus overcoming the membrane stiffness. It is obtained by pressurising the membrane in the air.

7.1 Introduction

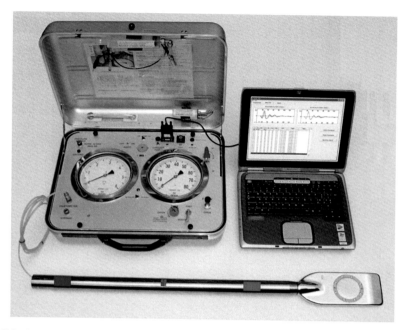

Fig. 7.2 Seismic dilatometer

In softer soils, the equations suggested by Schmertmann (1986) will give more realistic values. They are:

$$p_0 = 1.05(A + \Delta A - z_m) - 0.05(B - \Delta B - z_m) \tag{7.3}$$

$$p_1 = B - \Delta B - z_m \tag{7.4}$$

Here, z_m is the zero reading of the pressure gauge which is generally zero for new gauge. Corrected pressure p_2 is given approximately by

$$p_2 = C + \Delta A - z_m \tag{7.5}$$

Seismic flat dilatometer (Fig. 7.2) was introduced in 2006, with two geophones located above the blade, 500 mm apart. When a plate is struck at the ground level, the signals are received at different times by the geophones. These data are used to compute the shear wave velocity v_s with 1–2 % repeatability, and its variation with depth can be established. From the shear wave velocity, the small strain shear modulus G_0 can be computed as

$$G_0 = \rho V_s^2 \tag{7.6}$$

In addition, SDMT gives all other parameters derived from DMT.

Borehole dilatometer or rock dilatometer is different device. It is a radially expandable cylindrical probe that is inserted into the borehole and expanded by air at pressures up to 30 MPa. The device is used to assess the deformability of in situ rock mass. The correlations discussed in this chapter are for the flat dilatometers used in soils.

Mayne et al. (2009) categorised all the in situ tests as "old" and "new" methods, with seismic cone penetration tests with pore pressure measurements (SCPTu) and seismic dilatometer tests (SDMT) falling under the new methods. A DMT can be used in very soft to very stiff clays or marls, with $c_u = 2$–1000 kPa.

7.2 Intermediate DMT Parameters

Interpretation of dilatometer test data is essentially empirical, based on the two corrected pressures p_0 and p_1. Here p_0 is the corrected pressure A and p_1 is the corrected pressure B. The *three* intermediate parameters that are used for deriving the other soil parameters are the *material index I_D, horizontal stress index K_D*, and *dilatometer modulus E_D*. They are computed empirically.

Material index I_D is defined as:

$$I_D = \frac{p_1 - p_0}{p_0 - u_0} \qquad (7.7)$$

where, u_0 is the hydrostatic pore water pressure that was present prior to insertion of the blade. Material index, which is low for clays (<0.6), medium for silts (0.6 – 1.8) and high for sands (>1.8), is used to identify the soil. It is typically in a range of 0.1 to 10.

Horizontal stress index K_D is defined as:

$$K_D = \frac{p_0 - u_0}{\sigma'_{vo}} \qquad (7.8)$$

where, σ'_{vo} is the effective in situ overburden stress. K_D can be seen as the K_0 (coefficient of earth pressure at rest) amplified by the penetration (Marchetti 1994). It is typically about two in normally consolidated clays, which is significantly larger than K_0. In overconsolidated soils, K_D is greater than 2. Horizontal stress index is used to determine horizontal stress and hence K_0, OCR and undrained shear strength (c_u) in clays and effective friction angle (ϕ') in sands. K_D is a measure of the soil's resistance to volume reduction, and is highly sensitive to aging and is sometimes called the stress history index.

Dilatometer modulus E_D is obtained from elastic analysis by relating the membrane displacement s_0 to the pressure difference $\Delta p = p_0 - p_1$. Assuming an elastic half space surrounding the membrane, using the theory of elasticity, the displacement s_0 can be expressed as (Marchetti 1980):

$$s_0 = \frac{2D(p_1 - p_0)}{\pi} \frac{1 - \nu^2}{E} \qquad (7.9)$$

where $E =$ Young's modulus, $D =$ membrane diameter and $\nu =$ Poisson's ratio. For membrane diameter $D = 60$ mm and $s_0 = 1.1$ mm, Eq. (7.9) becomes

$$E_D = \frac{E}{1 - \nu^2} = 34.7(p_1 - p_0) \qquad (7.10)$$

Dilatometer modulus is used in determining the constrained modulus and hence modulus of elasticity. It must be noted that the soil is loaded laterally in determining the modulus. In reality, the soil modulus is required for vertical loading. The E_D computed from Eq. (7.10) is drained in sand, undrained in clays and is partially drained in sand-clay mixtures.

7.3 Correlations

One of the main uses of the dilatometer test is in identifying the soil type. Figure 7.3 shows the soil identification chart based on the material index I_D and the dilatometer modulus E_D. The original chart was proposed by Marchetti and Crapps (1981), which was modified slightly by Schmertmann (1986). The chart also gives an estimate of the unit weights.

Marchetti (1980) showed that for uncemented clays with $I_D < 1.2$ and sands/silts with $I_D \geq 1.2$

$$OCR = (0.5K_D)^{1.56} \qquad (7.11)$$

This was verified experimentally by Kamei and Iwasaki (1995) and theoretically by Finno (1993). Kulhawy and Mayne (1990) showed test data suggesting that the coefficient 0.5 in Eq. (7.11) can vary in the range of 0.27–0.75, depending on the degree of fissuring, sensitivity and geologic origin.

Noting that

$$\left(\frac{c_u}{\sigma'_{v0}}\right)_{OC} = \left(\frac{c_u}{\sigma'_{v0}}\right)_{NC} OCR^{0.8} \qquad (3.46)$$

and assuming $\left(\frac{c_u}{\sigma'_{v0}}\right)_{NC} = 0.22$ as suggested by Mesri (1975), Marchetti (1980) proposed that for $I_D \leq 1.2$,

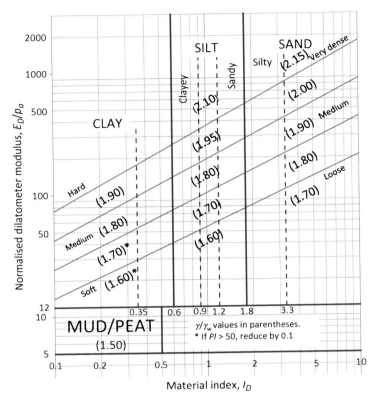

Fig. 7.3 Soil identification chart (After Schmertmann 1986)

$$\left(\frac{c_u}{\sigma'_{v0}}\right)_{OC} = \left(\frac{c_u}{\sigma'_{v0}}\right)_{NC}(0.5K_D)^{1.25} \approx 0.22(0.5K_D)^{1.25} \quad (7.12)$$

Kulhawy and Mayne (1990) suggest that this equation should be limited to $I_D \leq 0.6$.

The undrained shear strength of the clay can be determined from Eq. (7.12). Similar equations suggested by others include the following.

Iwasaki and Kamei (1994):

$$c_u = 0.118E_D \quad (7.13)$$

Kamei and Iwasaki (1995):

$$\left(\frac{c_u}{\sigma'_{v0}}\right)_{OC} \approx 0.35(0.47K_D)^{1.14} \quad (7.14)$$

Constrained Modulus M_{DMT}

The one-dimensional constrained modulus (D) can be determined from an oedometer test using Eq. (2.47).

$$D = \frac{1}{m_v} = \frac{1-\nu}{(1+\nu)(1-2\nu)}E \qquad (2.47)$$

In this chapter, the constrained modulus is denoted by M_{DMT}, when it is determined using a DMT. M_{DMT} can be derived from the dilatometer modulus E_D as:

$$M_{DMT} = R_M E_D \qquad (7.15)$$

where, R_M is a function of K_D and I_D. K_D typically varies in the range of 2–20. R_M varies in the range of 1–3. Typically, M_{DMT} varies in the range of 0.4 to 400 MPa.

$$R_M = 0.14 + 2.36 \log K_D \qquad \text{for } I_D \leq 0.6 \qquad (7.16)$$
$$R_M = R_{M,0} + (2.5 - R_{M,0}) \log K_D \qquad \text{for } 0.6 < I_D < 3 \qquad (7.17)$$
$$\text{with } R_{M,0} = 0.14 + 0.15(I_D - 0.6) \qquad (7.18)$$
$$R_M = 0.5 + 2.0 \log K_D \qquad \text{for } I_D \geq 3 \qquad (7.19)$$

If $K_D > 10$,

$$R_M = 0.32 + 2.18 \log K_D \qquad (7.20)$$

When computing M_{DMT} using Eq. (7.15), R_M should not be less than 0.85.

Young's Modulus E

From the definition of E_D, the Young's modulus can be computed as

$$E = (1 - \nu^2) E_D \qquad (7.21)$$

From Eq. (2.60), for $\nu = 0.0$ to 0.3, $E = (0.75 \text{ to } 1.0) D$. Assuming $D = M_{DMT}$, E can be estimated once M_{DMT} is determined.

Effective Friction Angle ϕ'

Marchetti (1997) suggested an approximate equation for the effective friction angle, which he believes is a lower bound value that underestimates the in situ friction angle by 2–4°. The equation is

$$\phi' \text{(degrees)} = 28 + 14.6 \log K_D - 2.1 (\log K_D)^2 \qquad (7.22)$$

From tests conducted in ML and SP-SM soils in Venice, Ricceri et al. (2002) suggested an upper bound as

Table 7.1 Suggested values of β_K (After Kulhawy and Mayne 1990)

Soil type	β_K
Fissured clays	0.9
Insensitive clays	1.5
Sensitive clays	2.0
Glacial till	2.0

$$\varphi' = 31 + \frac{K_D}{0.236 + 0.066 K_D} \quad (7.23)$$

Coefficient of Earth Pressure at Rest K_0

Marchetti (1980) proposed that the coefficient of earth pressure at rest K_0 can be estimated from the horizontal stress index K_D using the following equation, for clays with $I_D < 1.2$ and sands or silts with $I_D \geq 1.2$.

$$K_0 = \left(\frac{K_D}{1.5}\right)^{0.47} - 0.6 \quad (7.24)$$

Equation (7.24) was based on data from insensitive Italian clays and uncemented normally consolidated sands. This was modified by Kulhawy and Mayne (1990) as

$$K_0 = \left(\frac{K_D}{\beta_K}\right)^{0.47} - 0.6 \quad (7.25)$$

where β_K depends on the soil type and geologic origin. Some values suggested by Kulhawy and Mayne (1990) are shown in Table 7.1.

Powell and Uglow (1988) showed that for young UK clays

$$K_0 = 0.34 K_D^{0.55} \quad (7.26)$$

Settlement Computations

Assuming one-dimensional compression, settlements can be computed by

$$\text{Settlement} = \sum \frac{\Delta \sigma_v}{M_{DMT}} \Delta z \quad (7.27)$$

where M_{DMT} is the constrained modulus derived from DMT from Eq. (7.15). When the settlement is truly 3-dimensional, it may be computed from

$$\text{Settlement} = \sum \frac{1}{E}[\Delta \sigma_v - \nu(\Delta \sigma_x + \Delta \sigma_y)] \quad (7.28)$$

7.4 Summary

Dilatometer is one of the latest arrivals among the in situ testing devices. It has become more versatile with the development of seismic dilatometer, which also measures the shear wave velocity. The three major indices computed in the dilatometer test are the material index I_D, horizontal stress index K_D, and dilatometer modulus E_D. These three are used in identifying the soil type and determining the soil parameters.

References

Finno RJ (1993) Analytical interpretation of flat dilatometer penetration through saturated cohesive soils. Geotechnique 43(2):241–254

Iwasaki K, Kamei T (1994) Evaluation of in situ strength and deformation characteristics of soils using flat dilatometer. JSCE J Geotech Eng 499(III-28):167–176 (in Japanese)

Kamei T, Iwasaki K (1995) Evaluation of undrained shear strength of cohesive soils using a flat dilatometer. Soils Found 35(2):111–116

Kulhawy FH, Mayne PW (1990) Manual on estimating soil properties on foundation design, EL-6800. Electrical Power research Institute, California

Marchetti S (1980) In situ tests by flat dilatometer. J Geotech Eng Div ASCE 106(GT3):299–321

Marchetti S (1994) The flat dilatometer: design applications. In: Proceedings of the 3rd geotechnical engineering conference, Cairo University, Keynote lecture, 26 pp

Marchetti S (1997) The flat dilatometer design applications, Keynote lecture, 3rd geotechnical engineering conference, Cairo University, Egypt, pp 421–448

Marchetti S, Crapps DK (1981) Flat dilatometer manual, DMT operating manual. GPE Inc., Gainesville

Marchetti S, Monaco P, Totani G, Calabrese M (2001) The flat dilatometer test (DMT) in soil investigations. A report by the ISSMGE Committee TC16. In: Proc. In Situ 2001, Intnl. Conf. on in situ measurement of soil properties, Bali, Indonesia, 41 pp

Mayne PW, Coop MR, Springman SM, Huang AB, Zornberg JG (2009) State-of-the-art paper SoA-1, Geomaterial behaviour and testing. In: Proceedings of 17th ICSMGE, Alexandria, 2777–2872

Mesri G (1975) Discussion of New design procedure for stability of soft clays. J Geotech Eng Div ASCE 101:409–412

Powell JJM, Uglow IM (1988) The interpretation of Marchetti dilatometer test in UK clays. In: Proceedings of penetration testing in UK, ICE, University of Birmingham, pp 269–273

Ricceri G, Simonini P, Cola S (2002) Applicability of piezocone and dilatometer to characterise the soils of the Venice Lagoon. Geotech Geol Eng 20(2):89–121

Schmertmann JH (1986) Suggested method for performing flat dilatometer test. Geotech Test J ASTM 9(2):93–101

Chapter 8
Vane Shear Test

Abstract It describes the vane shear test which is one of the most widely used in situ tests to assess the undrained shear strength of saturated cohesive soils. The test equipment and procedure to carry out the test are described in detail as well as interpretation of the data, and how the results could be used in geotechnical design. Empirical correlations are provided to correct the measured field vane shear strength values and the shortcomings in these methods are highlighted.

Keywords Correlations • Vane shear • Sensitivity • Undrained shear strength • Over consolidation ratio

8.1 Vane Shear Test – General

The vane shear test is one of the most widely used in situ tests to assess the undrained strength of saturated cohesive materials. The test is most suited for materials having a consistency of soft to firm clayey soils, generally considered to be weak and compressible. The test is not suitable for cohesionless materials such as sand and gravel as undrained conditions cannot be maintained in such soils. The use of the vane shear test in fibrous peat is also questionable.

The test is understood to have been originated in Sweden in the early twentieth century but became popular more towards the 1940s (see Flodin and Broms 1981). An overview of the vane shear test has been presented by Walker (1983). It is a simple and easy test and can be performed by advancing to the depth where the test has to be conducted, including from the base of a borehole. The test has several advantages including simplicity, robustness and short test duration. The test appears to create less disturbance compared to many other in situ and laboratory tests.

The test indirectly provides the assessment of the over consolidation ratio of the soil deposit based on empirical rules. The vane shear test allows the measurement of the peak and residual strength and therefore the sensitivity of cohesive soils, which is not possible by any other in situ test.

As presented in Table 1.1, Lunne et al. (1997) summarized the various in situ tests in operation at the time and classified them according to their applicability and usefulness in deriving different design parameters as well as in different material types. The table indicates that only the vane shear test has a high applicability classification compared to other in situ tests listed to obtain the undrained shear strength of soil (clay). There have been advances since the publication of this table, especially piezocone and dilatometer testing, which are covered in Chaps. 5 and 7.

Although the vane shear test has its advantages, there has also been research that highlights issues related to parameters derived from vane shear testing. Readers are referred to a useful summary by Johnston (1983) and Donald et al. (1977) who did a critical evaluation of the vane shear test. With all the advantages and disadvantages listed, the vane shear test still appears to be one of the most popular amongst practitioners.

8.2 Vane Shear Test Equipment and Procedure in the Field

Vane shear equipment consists of two thin vanes perpendicular to each other (cruciform) connected to a solid pushing rod (see Figs. 8.1 and 8.2). The test comprises inserting the vane to the required depth and rotating about a vertical axis which allows the soil to shear. The test is carried out generally every 0.5 m or 1 m depth or at depths selected by the designer based on other available data such as information on the geology or continuous profile provided by a cone penetrometer or a piezocone test.

The test can be either done at the base of a borehole or directly in the ground. The test procedure could be summarized as follows based on ASTM D2573-08 (See Fig. 8.3):

1. Position equipment over test location. The test can be performed in a pre-drilled borehole (pushing from the surface) or by drilling through a vane housing. If a pre-drilled hole is used, predrilling should cease such that the vane tip can penetrate undisturbed soil for a depth at least $5 \times$ outside diameter of the hole to reduce the effect of ground disturbance and edge effects. If a vane housing (Fig. 8.4) is used, advance it to a depth at $5 \times$ housing diameter above the nominated test depth.
2. Push down the vane slowly with a single thrust from the bottom of the borehole or vane housing to the nominated test depth.
3. Within 5 min, apply a torque and slowly and continuously rotate the vane at a rate of 0.1 deg/s with permissible variation of 0.05–0.2 deg/s. and record the maximum torque registered (which represents the peak strength of the soil). It is suggested that readings be recoded every 15 s.
4. Rotate the vane rapidly, at least 5–10 revolutions, and record the torque which represents the remoulded (i.e., residual) strength.

8.2 Vane Shear Test Equipment and Procedure in the Field

Fig. 8.1 Typical geometry of a field vane

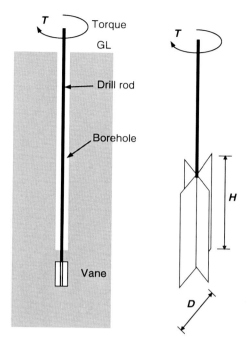

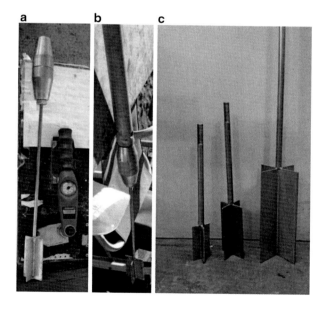

Fig. 8.2 Photograph of a field vane apparatus (**a**) Vane and torque meter (**b**) Vane connected to drill rod (**c**) Different size vanes (Courtesy Allan McConnell, IGS)

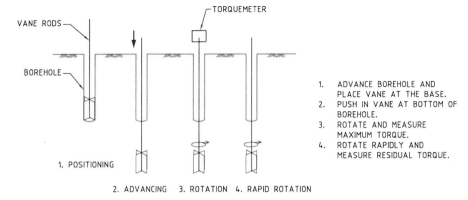

Fig. 8.3 Vane shear test operation in a borehole

Fig. 8.4 Vane housing (Courtesy Allan McConnell, IGS)

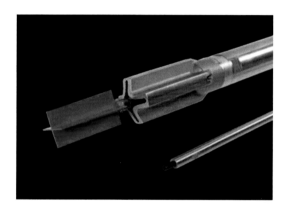

8.3 Assessment of Shear Strength in the Field Using the Vane Shear Test

The applied torque is resisted by the shear stress mobilized along the failure surface. Since the test is carried out relatively fast, undrained conditions can be assumed and hence the shear stress at failure is the same as the undrained shear strength, c_u. Based on the schematic diagram shown in Fig. 8.1, the maximum torque T required to rotate the vane shear blades and cause failure could be expressed as per Eq. (8.1):

$$T = M_{top} + M_{base} + M_{side} \qquad (8.1)$$

where

M_{top} = Resisting moment at the top of the blades/cylinder
M_{base} = Resisting moment at the base of the blades/cylinder
M_{side} = Resisting moment at the sides of the cylinder

8.3 Assessment of Shear Strength in the Field Using the Vane Shear Test

By taking moments about the shaft axis:

$$M_{side} = (\pi D H) \times c_u \times \frac{D}{2} \tag{8.2}$$

and

$$M_{top} = M_{base} = \int_0^{\frac{D}{2}} (2\pi r \, dr) \times c_u \times r \tag{8.3}$$

where

D = Diameter of the cylinder (i.e. width of the vane blade)
H = Height of vane
c_u = Undrained shear strength
r = Radius of the circular element of thickness dr.

Combining Eqs. (8.1), (8.2) and (8.3):

$$T = \left[(\pi D H) \times c_u \times \frac{D}{2}\right] + 2 \times \int_0^{\frac{D}{2}} (2\pi r \, dr) \times c_u \times r \tag{8.4}$$

which simplifies to:

$$T = \left[\frac{D^2 H}{2} + \frac{D^3}{6}\right] \times \pi c_u \tag{8.5}$$

As highlighted previously, the H/D ratio is usually 2 and Eq. (8.5) simplifies to the following:

$$T = \frac{7}{6} \pi D^3 \times c_u \tag{8.6}$$

or

$$c_u = T / \left(\frac{7}{6} \pi D^3\right) \tag{8.7}$$

or

$$c_u = T / (3.67 D^3) \tag{8.8}$$

The peak undrained shear strength is calculated from Eq. (8.8) using the maximum torque recorded by the test. The remoulded shear strength is also assessed using the same equation but, in this instance, using the torque measured towards the end of the test when five to ten revolutions are done very rapidly (Step 4 in Sect. 8.2).

8.3.1 Assessment of Sensitivity of Clay

As previously mentioned, the vane shear test could be used to assess the sensitivity of clayey soils. Sensitivity is a measure of the loss of undrained soil strength when disturbed and is defined as follows:

$$Sensitivity = \frac{peak\ undisturbed\ shear\ strength}{remoulded\ shear\ strength} \tag{8.9}$$

As discussed by Mitchell and Houston (1969), it was Terzaghi (1944) who proposed the above definition for strengths determined from unconfined compression test but which is not useful for highly sensitive clays because unconfined compression test specimens cannot be formed as the remoulded strength is so low. Mitchell and Houston (1969) provide Table 8.1 summarizing several classifications proposed by different authors. Section 3.3.7 provides additional sensitivity definitions adopted by US, Canada and Sweden.

8.4 Vane Shear Test Corrections

No in situ or laboratory test is perfect and the vane shear test is no different (Sect. 8.1). Bjerrum (1972) carried out back calculation of several embankment failures and found that factors of safety of the failed embankments were significantly higher than 1.0. He concluded that plasticity of soil has a major influence and should be corrected for, prior to using the undrained shear strength values derived from vane in the design of embankment loading and excavation stability. It was suggested that the c_u derived in the field be multiplied by a correction factor μ. The correction factor, μ, proposed by Bjerrum (1972) to multiply the measured field vane shear strength (c_{uFV}) in order to obtain the mobilized shear strength (c_u) is related to Plasticity Index (*PI*) and this relationship is shown in Fig. 8.5. Figure 8.5 also shows relationships proposed based on research by others including Morris and Williams (1994) and Chandler (1988). Although data in their research show some scatter, the plots indicate that the shear strength correction shows a trend that decreases with plasticity of the soil. Ladd et al. (1977) included data from other sites and they found that the scatter increases. As Ladd et al. (1977) point out, *"These additional points increase the scatter about Bjerrum's recommended curve, which has led to some (e.g. Milligan 1972; Schmertmann 1975) to seriously question this entire design approach in view of the scatter, probable variations in*

8.4 Vane Shear Test Corrections

Table 8.1 Sensitivity classification

Skempton and Northey (1952)	Rosenqvist (1955)	Shannon and Wilson (1964)
~1.0: insensitive clays	~1.0: insensitive clays	<3: Low
1–2: clays of low sensitivity	1–2: slightly sensitive clays	3–5: Low to medium
2–4: clays of medium sensitivity	2–4: medium sensitive clays	5–7: Medium
4–8: sensitive clays	4–8: very sensitive clays	7–11: Medium to high
>8: extra-sensitive clays	8–16: slightly quick clays	11–14: high
>16: quick clays	16–32: medium quick clays	14–20: High to very high
	32–64: very quick clays	20–40: Very high
	>64: extra quick clays	>40: Extremely high

After Mitchell and Houston (1969)

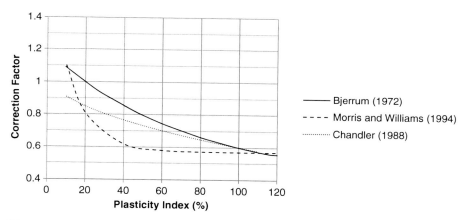

Fig. 8.5 Field vane correction factor vs plasticity index

the determination of the plasticity index (especially with varved clays) and uncertainties in the circular arc stability analyses used to develop the correction. Variable field vane test procedures should be added to this list. Nevertheless, the trend of the relationship is considered sound and supported by the data, which are distributed fairly evenly about Bjerrum's curve and generally fall within 20 % of its value. ………".

Since Bjerrum's work, several others have examined the correction factor phenomenon. Leroueil et al. (1990) and Leroueil (2001) provide information for soft slays suggesting that no correction is necessary. Azzouz et al. (1983) suggest that the Bjerrum correction factor should be further reduced by about 10 % for field situations that resemble plane strain conditions. Morris and Williams (1994) proposed the following (See Fig. 8.5):

$$\mu = 1.18 \, e^{-0.08 \, PI} + 0.57 \, (\text{for} \, PI > 5) \tag{8.10}$$

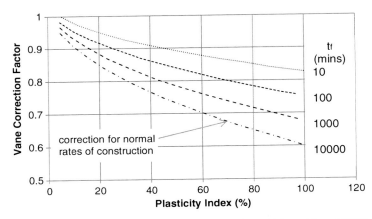

Fig. 8.6 Vane correction factor in terms of PI and time to failure (Adopted from FHWA NHI-01-031 – Mayne et al. 2001 and Chandler 1988)

$$\mu = 7.01\ e^{-0.08(LL)} + 0.57\ (\text{for } LL > 20) \qquad (8.11)$$

where

LL = Liquid Limit (%)
PI = Plasticity Index (%)

ASTM D2573-08, while not endorsing or recommending any method for correction, cites the following correction factor by Chandler (1988) in an Appendix (non-mandatory information):

$$\mu = 1.05 - b(PI)^{0.5} \qquad (8.12)$$

where parameter 'b' is a rate factor that depends on the time to failure (t_f in minutes) in the actual failure (not in the field test) and given by:

$$b = 0.015 + 0.0075 \log t_f \qquad (8.13)$$

The combined relationship is given in Fig. 8.6. ASTM D2573-08 states that, for guidance, for embankments on soft ground, t_f is of the order of 10^4 min and Eqs. (8.12) and (8.13) become:

$$\mu = 1.05 - 0.045\ (PI)^{0.5} \qquad (8.14)$$

Although various equations and relationships have been proposed over the years, Bjerrum's correction (or curve) is still widely accepted by practicing engineers and remains very popular to this day.

8.5 Correlations for c_u in Normally Consolidated Soils

Some correlations for the undrained shear strength were discussed in Chap. 2. However, for completeness and their importance and relevance to vane shear test some of the correlations are revisited in this Chapter.

From early research it was quite evident that the undrained shear strength is related to soil plasticity as well as the effective vertical stress. Skempton (1948) proposed Eq. (8.15), a widely used empirical equation, which relates the ratio of undrained shear strength measured by the field vane (c_u $_{FV}$) and effective vertical overburden stress (σ_v') to the plasticity index (PI) for normally consolidated (NC) soils i.e. over consolidation ratio (OCR) = 1.0.

$$\left(c_{u\ FV}/\sigma_v'\right)_{NC} = 0.11 + 0.0.0037\,PI \tag{8.15}$$

This relationship indicates how the undrained strength ratio derived from the vane shear test increases with PI for normally consolidated soils. Therefore, if the physical characteristics (Atterberg limits) of a normally consolidated soil deposit are known the undrained shear strength obtained from a field vane test could be estimated. The values derived should be subjected to the correction factor discussed previously to obtain a mobilized undrained shear strength expected.

By combining the variation of the shear vane strength with PI and vane correction factor with PI, Mesri (1975) proposed that the mobilized undrained strength ratio (c_u/σ_v')$_{NC}$ for NC and slightly OC soils, is independent of PI and is a constant of 0.22 (±0.03):

$$\left(c_u/\sigma_v'\right)_{NC} = 0.22 \pm 0.03 \tag{8.16}$$

A constant value (or nearly constant value) for the mobilized strength ratio is not surprising when one considers the opposite effects created by:

(a) the increasing field vane shear strength ratio with increasing PI (Eq. (8.15)) and
(b) the vane correction factor (μ) decreasing with increasing PI (Fig. 8.5).

Chandler (1988), based on examination of other data suggests that the overall range of scatter in the value of Mesri's quoted mobilized strength ratio is of the order of (±0.05):

$$\left(c_u/\sigma_v'\right)_{NC} = 0.22 \pm 0.05 \tag{8.17}$$

According to Jamiolkowski et al. (1985), Larsson (1980) suggested a similar equation for clays (PI < 60 %) after examining several embankment failures:

$$\left(c_u/\sigma_v'\right)_{NC'} = 0.23 \pm 0.04 \tag{8.18}$$

8.6 Correlations for c_u in Over Consolidated Soils

The general equation for undrained strength ratio for an overconsolidated soil can be approximated by (Jamiolkowski et al. 1985; Ladd and DeGroot 2003) as follows:

$$(c_{uFV}/\sigma_{v'})_{OC} = (c_{uFV}/\sigma_{v'})_{NC}(OCR)^m \quad (8.19)$$

Chandler (1988) adopted an 'm' value of 0.95 whereas Ladd and DeGroot (2003) obtained a value of 0.89 ± 0.08 for the data analysed by them. Chandler (1988) summarized the parameter $(c_{uFV}/\sigma_{v'})_{NC}$ and 'm' given by Jamiolkowski et al. (1985) for field vane tests and this is shown in Table 8.2.

It is generally accepted that the 'm' value varies with the type of test. A value of 0.8 is used quite often by practicing engineers for 'm', although it was first proposed by Ladd et al. (1977) based on CK$_o$U direct simple shear tests and not vane shear tests. Then Eq. (8.19) becomes:

$$(c_{uFV}/\sigma_{v'})_{OC} = (c_{uFV}/\sigma_{v'})_{NC}(OCR)^{0.8} \quad (8.20)$$

To calculate mobilized undrained shear strength (c_u), Bjerrum's vane shear correction needs to be applied.

$$(c_u/\sigma_{v'})_{OC} = \mu (c_{uFV}/\sigma_{v'})_{OC} = \mu (c_{uFV}/\sigma_{v'})_{NC}(OCR)^{0.8} \quad (8.21)$$

If a field vane shear test is carried out in the field, the results could then be used to calculate the OCR of the soil deposit by combining a relationship from Fig. 8.6 with Eqs. (8.15) and (8.21) if the Atterberg limits of the soils are known.

Another approach to obtain OCR would be the use of Mesri's Eq. (8.16) if expressed using the effective maximum past pressure (σ'_p) for an overconsolidated soil, i.e.,

$$(c_u/\sigma'_p)_{OC} = 0.22 \pm 0.03 \quad (8.22)$$

Therefore, if a vane shear test is carried out and the assessed strength is corrected to obtain c_u (see Sect. 8.4), the maximum past pressure, σ'_p, can be easily calculated.

Table 8.2 $(c_{uFV}/\sigma_{v'})_{NC}$ and m values

Parameter	$(c_{uFV}/\sigma_{v'})_{NC}$	m
Typical range of values (all sites)	0.16–0.33	0.80–1.35
Extreme value (one in each case)	0.74	1.51
Mean (all values)	0.28	1.03
Mean (discarding extreme value)	0.22	0.97

8.6 Correlations for c_u in Over Consolidated Soils

As the vertical effective stress at the test depth can be calculated to provide σ'_v, OCR can be calculated as:

$$OCR = \sigma'_p/\sigma'_v \qquad (8.23)$$

Several authors have commented on the derived undrained strength ratio for organic soils. Jamiolkowski et al. (1985) concludes that *"the in situ c_u/σ'_p appropriate for analyses of embankment stability probably falls within a fairly narrow range for most soft sedimentary clays of moderate to low plasticity. .... However, there is some evidence indicating that c_u/σ'_p for high plastic organic clays is higher than quoted above. ..... For example, the ratio for non-fiberous peats is very high."*

Terzaghi et al. (1996) state that the *"fiber content of organic soils may act as a localized reinforcement or drainage veins across the relatively thin shear zone in the vane test and lead to vane strengths that are too high. Therefore Eq. (8.23) may underestimate the mobilized undrained shear strength for organic clays and silts. It suggests, for organic soils (excluding peats), an additional reduction factor of 0.85 should be used in addition to the vane correction factor in Eq. (8.10), and $\left(c_u/\sigma'_p\right)_{OC} = 0.26$ should be used."*

Ladd (1991) recommends procedures to obtain c_u/σ'_p vs OCR relationships but suggests there are three levels of assessments:

Level A – For final design of major projects and for sites with soils exhibiting strength anisotropy or unusual features (such as organic soils, fissuring etc) and projects where lateral deformation is important

Level B – For preliminary design and for final design of less important projects involving "ordinary" soils with low to moderate anisotropy

Level C – For preliminary feasibility studies and to check the reasonableness of in situ or laboratory test assessments

Ladd (1991) suggests that empirical correlations play a role in Level C selection of strength parameters in the form of

$$\left(c_u/\sigma_{v'}\right)_{OC} = S(OCR)^m \qquad (8.24)$$

where S is obtained by substituting $m = 1$ in the above equation, i.e.,

$$S = \left(c_u/\sigma'_v\right)_{NC} \qquad (8.25)$$

Ladd (1991) states that when estimating 'm', which according to Modified Cam-Clay theory of soil behavior, should equal to $1 - C_s/C_c$ (relate to the slopes of the swelling and virgin compression lines, respectively – Roscoe and Burland 1968). Based on his interpretation of data and experience, Ladd and DeGroot (2003) recommended values for "S" and "m" in Eq. 8.24 as presented in Table 8.3.

Table 8.3 S and m values for estimation of average c_u via SHANSEP equation (Ladd and DeGroot 2003)

Soil Description	S	m^a	Remarks
1. Sensitive cemented marine clays ($PI < 30\ \%, LI > 1.5$)	0.20 Nominal SD = 0.015	1.00	Champlain clays of Canada
2. Homogeneous CL and CH sedimentary clays of low to moderate sensitivity ($PI = 20$–$80\ \%$)	$S = 0.20 + 0.05$ ($PI/100$) or simply 0.22	$0.88(1 - C_s/C_c) \pm 0.06$ SD or simply 0.8	No shells or sand lenses-layers
3. Northeastern US varved clays	0.16	0.75	Assumes that Direct Simple Shear (DSS) mode of failure predominates
4. Sedimentary deposits of silts and organic soils (Atterberg Limits plot below A-line) and clays with shells	0.25 Nominal SD = 0.05	$0.88(1 - C_s/C_c) \pm 0.06$ SD or simply 0.8	Excludes peat

$^a m = 0.88(1 - Cs/Cc)$ based on analysis of CK0U DSS data on 13 soils with max. $OCR = 5$–10
SD Standard deviation, *LI* Liquidity Index, *CL/CH* Low/High Plasticity, C_s and C_c Slopes of the swelling and virgin compression lines (Roscoe and Burland 1968)

8.7 Summary

This chapter provides a brief overview of the vane shear test, the procedure, interpretation of the data, and how the results could be used in geotechnical design. Empirical correlations are provided to correct the measured field vane shear strength values and the shortcomings in these methods are highlighted.

References

ASTM D2573-08 Standard test method for field vane shear test in cohesive soil
Azzouz AS, Baligh MM, Ladd CC (1983) Corrected field vane strength for embankment design. J Geotech Eng ASCE 109(5):730–734
Bjerrum L (1972) Embankments on soft ground. In: Proceedings Performance of earth and earth-supported structures, vol 2. ASCE, Lafayette, Ind., pp 1–54
Chandler RJ (1988) The in-situ measurement of the undrained shear strength of clays using the field vane. Vane shear strength testing of soils: field & lab studies, STP1014, ASTM, West Conshohocken, pp 13–44
Donald IB, Jordan DO, Parker RJ, Toh CT (1977) The vane test - a critical appraisal. In: Proceedings of 9th international conference on soil mechanics and foundation engineering, Tokyo, vol 1, pp 151–171
Flodin N, Broms B (1981) Historical development of civil engineering in soft clay. In: Brand EW, Brenner RP (eds) Soft clay engineering. Elsevier Publications, Amsterdam
Jamiolkowski M, Ladd CC, Germaine JT, Lancellotta R (1985) New developments in field and laboratory testing of soils. In: Proceedings of 11th international conference on soil mechanics and foundation engineering, vol 1, pp 57–153

References

Johnston IW (1983) Why in-situ testing? In: Ervin MC (ed) Proceedings of an extension course on in-situ testing for geotechnical investigation, Sydney. A A Balkema, Rotterdam

Ladd CC (1991) Stability evaluation during staged construction (22nd Terzaghi Lecture). J Geotech Eng 117(4):540–615

Ladd CC, DeGroot DJ (2003) Recommended practice for soft ground site characterization (Arthur Casagrande Lecture). In: Proceedings 12th Pan American conference on soil mechanics and geotechnical engineering, Massachusetts Institute of Technology, Cambridge, MA

Ladd CC, Foott R, Ishihara K, Schlosser F, Poulos HG (1977) Stress-deformation and strength characteristics. In: Proceedings 9th international conference on soil mechanics and foundation engineering, Tokyo, 2, pp 421–494

Larsson R (1980) Undrained shear strength in stability calculations of embankments & foundations on clays. Can Geotech J 17(4):591–602

Leroueil S (2001) Natural slopes and cuts: movement and failure mechanisms, 39th Rankine Lecture. Geotechnique 51(3):197–243

Leroueil S, La Rochelle P, Tavenas F, Roy M (1990) Remarks on the stability of temporary cuts. Can Geotech J 27(5):687–692

Lunne T, Robertson PK, Powell JJM (1997) Cone penetration testing in geotechnical practice. Blackie Academic and Professional, London, 312pp

Mayne PW, Christopher BR, DeJong J (2001) Manual on subsurface investigations, National highway institute publication no 6. FHWA NHI-01-031 Federal Highway Administration, Washington, DC

Mesri G (1975) Discussion on "New design procedure for stability of soft clays". J Geotech Eng Div ASCE 101(GT4):409–412

Milligan V (1972) Panel discussion, Session I. Embankments on soft ground. In: Proceedings ASCE specialty conference on performance of earth and earth-supported structures, Lafayette, vol III, pp 41–48

Mitchell JK, Houston WN (1969) Causes of clay sensitivity. J Soil Mech Found Eng ASCE 95(SM3):845–871

Morris PM, Williams DJ (1994) Effective stress vane shear strength correction factor correlations. Can Geotech J 31(3):335–342

Roscoe KH, Burland JB (1968) On the generalized stress-strain behavior of "wet" clay. In: Heyman J, Leckie FA (eds) Engineering plasticity. Cambridge University Press, Cambridge, pp 535–609

Rosenqvist I (1955) Investigations in the clay electrolyte water system, Norwegian Geotechnical Institute Publication No 9, Oslo

Schmertmann JH (1975) Measurements of in situ shear strength. In: Proceedings ASCE specialty conference on in-situ measurement of soil properties, Raleigh NC, vol 2, pp 57–138

Shannon and Wilson, Inc. (1964) Report on anchorage area soil studies, Alaska to US Army Engineering District, Anchorage, Alaska, Seattle, Washington

Skempton AW (1948) Vane tests in the alluvial plain of River Forth near Grangemouth. Geotechnique 1(2):111–124

Skempton AW, Northey RD (1952) The sensitivity of clays. Geotechnique III(I):30–53

Terzaghi K (1944) Ends and means in soil mechanics. Eng J Can 27:608

Terzaghi K, Peck RB, Mesri G (1996) Soil mechanics in engineering practice, 3rd edn. John Wiley, New York

Walker BF (1983) Vane shear strength testing. In: Ervin MC (ed) In situ testing for geotechnical investigations. A A Balkema, Rotterdam, pp 65–72

Chapter 9
Laboratory Rock Tests

Abstract The rock mass is made of intact rock and one or more sets of discontinuities. The intact rock specimens that are recovered from coring represent a relatively small volume of the rock mass and do not fully reflect the presence of discontinuities. The behaviour of the rock mass is governed mostly by the discontinuities than the properties of the intact rocks. Uniaxial compressive strength, Brazilian indirect tensile strength and the point load strength are some of the properties that reflect the strength of the intact rock. Typical values of these parameters and the interrelationships among them are discussed in this chapter.

Keywords Correlations • Rock • Uniaxial compressive strength (UCS)

It is difficult to exclude rock mechanics completely from the day-to-day life of a geotechnical engineer. In many geotechnical projects, there is a need to deal with rocks and some understanding of their properties and the correlations that can be used to estimate them can become very valuable. While the emphasis throughout this book is on the parameters and correlations for soils, this very last chapter is devoted to rocks.

The classification and characterisation of the rock are generally carried out through (a) *intact rock* specimens tested in the laboratory and (b) *rock mass* tested in situ. When testing intact rock specimens, the most common parameters derived are as follows.

(a) Unit weight, water content, porosity and water absorption
(b) Hardness
(c) Durability
(d) Point load strength index
(e) UCS and stiffness
(f) Wave velocities
(g) Permeability

The intact rock parameters used in rock engineering designs include cohesion, friction angle, UCS, Young's modulus, shear modulus, bulk modulus, Poisson's ratio, shear strength and stiffness of discontinuities and tensile strength. The parameters derived from the intact specimens have to be translated to the larger rock mass with due considerations to the discontinuities present.

The measurements carried out in situ rock mass include deformability, strength, permeability and the in situ stresses. Some of the common in situ tests carried out in rocks are as follows.

(a) Borehole dilatometer test
(b) Borehole jack test
(c) Plate load test
(d) In situ direct shear test
(e) Borehole slotter test
(f) Hydraulic fracture test

These days geophysical tests are increasingly becoming popular for covering a larger terrain in a relatively shorter time. The geophysical tests include seismic refraction, seismic reflection, Spectral Analysis of Surface Waves (SASW), Multi-channel Analysis of Surface Waves (MASW), Ground Penetrating Radar (GPR), electrical resistivity tests, cross hole and down hole tests.

9.1 Rock Cores and RQD

The behaviour of a *rock mass* is governed by the presence of discontinuities (e.g., joints, fissures, faults), their orientations, strength etc. *Intact rock* is the material between the discontinuities, a specimen of which is tested in the laboratory. It requires good judgment in arriving at the bigger picture using the lab data and the discontinuities present in the field. We generally test the intact rock in the laboratory, and then extrapolate to the rock mass in the field situation, considering the presence of discontinuities, boundary conditions, etc.

The rock cores recovered from the boreholes are generally taken to the laboratory for strength and deformability tests. The common core size designations and their nominal diameters are given in Table 9.1. The core barrel may consist of single, double or triple tubes to minimize the disturbance. When attempting to obtain a rock core over a certain depth, due to the presence of joints and fractures, a significant length may be "lost". This can be seen as a measure of the quality of the intact rock. Two similar parameters commonly used to ascertain the quality of intact rock based on the drill record are *core recovery ratio* (CR) and *rock quality designation* (RQD). Core recovery ratio is defined as:

$$CR\ (\%) = \frac{\text{Length of rock core recovered}}{\text{Total length of the core run}} \times 100 \qquad (9.1)$$

Rock quality designation (*RQD*) is a modified measure of core recovery, defined as (Deere 1964):

Table 9.1 Core size designations and nominal diameters

Symbol	Nominal core diameter (mm)	(inches)
AQ	27.0	1–1/16
BQ	36.5	1–7/16
NQ	47.6	1–7/8
HQ	63.5	2–1/2
PQ	85.0	3–11/32
EX	22.2	7/8
AX	30.2	1–3/16
BX	41.3	1–5/8
NX	54.0	2–1/8

$$RQD\ (\%) = \frac{\sum \text{lengths of core pieces longer than 100 mm}}{\text{Total length of the core run}} \times 100 \quad (9.2)$$

Table 9.2 summarises the classification of rocks based on RQD, and the allowable bearing capacities suggested by Peck et al. (1974).

9.2 Permeability

Permeability tests on intact rock specimens are carried out in laboratories. Due to the presence of joints and fissures, known as discontinuities, the permeability of the rock mass (*secondary permeability*) in the field can be substantially greater than that of the intact rock specimen (*primary permeability*) in the laboratory. The flow characteristics is the rock mass are often governed by the flow through these discontinuities than the flow through the intact rock. Goodman (1980) tabulated some values of permeabilities as measured in the laboratory and in situ, where the in situ values were orders of magnitude larger. Some typical permeability values of different rock types, measured on intact rock specimens in the laboratory are given in Table 9.3 (Serafim 1968; Serafim and Del Campo 1965). Intact basalt and granite are typical of low permeability rocks, and intact sandstone and limestone generally have high permeabilities.

9.3 Uniaxial Compressive Strength

Uniaxial compression test (ASTM D7012; ISRM 1979a; AS 4133.4.2.1), also known as *unconfined compression test*, is the most common rock test for assessing the strength of intact rock and rock masses. It is generally carried out on specimens with diameters larger than NX core size (54 mm diameter) and length to diameter ratios of 2–3. In the literature, uniaxial compressive strength or unconfined

Table 9.2 RQD, in situ rock quality description and allowable bearing pressure

RQD (%)	Rock quality	Allowable bearing pressure (MPa)
0–25	Very poor	1–3
25–50	Poor	3–6.5
50–75	Fair	6.5–12
75–90	Good	12–20
90–100	Excellent	20–30

Peck et al. (1974)

Table 9.3 Typical values of permeability for intact rock specimens from laboratory tests

Rock type	Permeability (cm/s)
Basalt	1.0×10^{-12}
Breccia	4.6×10^{-10}
Calcite	0.7 to 93×10^{-9}
Dolertite	1 to 100×10^{-7}
Dolomite	4.6 to 12×10^{-9}
Gabbro	1 to 100×10^{-7}
Granite	5 to 20×10^{-11}
Limestone	7 to 1200×10^{-10}
Marble	1 to 10×10^{-5}
Mudstone, hard	6 to 20×10^{-7}
Sandstone	1.6 to 120×10^{-7}
Schist, black, fissured	1 to 3×10^{-4}
Slate	0.7 to 1.6×10^{-10}
Tuff	2.3×10^{-8}

Serafim (1968) and Serafim and Del Campo (1965)

compressive strength is denoted by σ_c, q_u or UCS. It is the most commonly used parameter in rock characterisation and designs, and numerical modelling of rock mechanics problems.

During the UCS test, it is possible to measure the Poisson's ratio and the Young's modulus. Typical values of Poisson's ratios of different rock types are given in Table 9.4. The values suggested by Lambe and Whitman (1979) are within a relatively narrower range compared to those suggested by Gercek (2007). The relative ductility of the rock can be classified based on the axial strain at peak load, as suggested in Table 9.5. Classification of rocks based on UCS, as suggested by ISRM (1978a) and Hoek and Brown (1997), is given in Table 9.6. The Canadian Foundation Engineering Manual suggests the presumed bearing capacity values shown in Table 9.7 for footings founded on rock. Zhang and Einstein (1998) suggested that the ultimate skin friction (f_s) of piles in rocks can be estimated as

$$f_s(\text{MPa}) = a(\sigma_c)^b \qquad (9.3)$$

where $a = 0.2$ to 0.3 and $b = 0.5$. The value of a increases with the socket roughness and Seidel and Haberfield (1994) reported a range of 0.22–0.67. The range

Table 9.4 Typical values of Poisson's ratios for rocks

Rock type	Poisson's ratio Gercek (2007)	Lambe and Whitman (1979)
Amphibolite		0.28–0.30
Andesite	0.20–0.35	
Anhydrite		0.30
Basalt	0.10–0.35	
Conglomerate	0.10–0.40	
Diabase	0.10–0.28	0.27–0.30
Diorite	0.20–0.30	0.26–0.29
Dolerite	0.15–0.35	
Dolomite	0.10–0.35	0.30
Dunite		0.26–0.28
Feldspathic Gneiss		0.15–0.20
Gabbro		0.27–0.31
Gneiss	0.10–0.30	
Granite	0.10–0.33	0.23–0.27
Granodiorite	0.15–0.25	
Greywacke	0.08–0.23	
Limestone	0.10–0.33	0.27–0.30
Marble	0.15–0.30	0.27–0.30
Marl	0.13–0.33	
Mica Schist		0.15–0.20
Norite	0.20–0.25	
Obsidian		0.12–0.18
Oligoclasite		0.29
Quartzite	0.10–0.33	0.12–0.15
Rock salt	0.05–0.30	0.25
Sandstone	0.05–0.40	
Shale	0.05–0.32	
Siltstone	0.05–0.35	
Slate		0.15–0.20
Tuff	0.10–0.28	

Gercek (2007) and Lambe and Whitman (1979)

Table 9.5 Relative ductility based on axial strain at peak load

Classification	Axial strain (%)
Very brittle	<1
Brittle	1–5
Moderately brittle (Transitional)	2–8
Moderately ductile	5–10
Ductile	>10

Handin (1966)

Table 9.6 Classification of soil and rock strengths

Grade	Description	Field identification	σ_c or q_u (MPa)	Rock types
S1	Very soft clay	Easily penetrated several inches by fist.	<0.025	
S2	Soft clay	Easily penetrated several inches by thumb.	0.025–0.05	
S3	Firm clay	Can be penetrated several inches by thumb with moderate effort.	0.05–0.10	
S4	Stiff clay	Readily indented by thumb, but penetrated only with great effort.	0.10–0.25[a]	
S5	Very stiff clay	Readily indented by thumbnail.	0.25[a]– 0.50[a]	
S6	Hard clay	Indented with difficulty by thumbnail.	>0.5[a]	
R0	Extremely weak rock	Indented by thumb nail.	0.25–1.0	Stiff fault gouge
R1	Weak rock	Crumbles under firm blows with point of geological hammer; Can be peeled by pocket knife.	1–5	Highly weathered or altered rock
R2	Weak rock	Can be peeled by a pocket knife with difficulty; Shallow indentations made by firm blow with a point of geological hammer.	5–25	Chalk, rock salt, potash
R3	Medium strong rock	Cannot be scraped or peeled with a pocket knife; Specimen can be fractured with a single firm blow of a geological hammer.	25–50	Claystone, coal, concrete, schist, shale, siltstone
R4	Strong rock	Specimen requires more than one blow by geological hammer to fracture it.	50–100	Limestone, marble, phyllite, sandstone, schist, shale
R5	Very strong rock	Specimen requires many blows of geological hammer to fracture it.	100–250	Amphibiolite, sandstone, basalt, gabbro, gneiss, granodiorite, limestone, marble, rhyolite, tuff
R6	Extremely strong rock	Specimen can only be chipped by a geological hammer.	>250	Fresh basalt, chert, diabase, gneiss, granite, quartzite

ISRM (1978a) and Hoek and Brown (1997)
[a]Slightly different to classification in geotechnical context

suggested by Zhang and Einstein (1998) leads to a conservative estimate of f_s. They also suggested an expression for estimating the ultimate end bearing pressure in rocks as

9.3 Uniaxial Compressive Strength

Table 9.7 Presumed bearing capacity values

Rock type and condition	Strength	Presumed allowable bearing capacity (kPa)	Remarks
Massive igneous and metamorphic rocks (e.g. granite, diorite, basalt, gneiss) in sound condition	High–Very high	10,000	Based on assumptions that the foundations are carried down to unweathered rock
Foliated metamorphic rocks (e.g. slate, schist) in sound condition	Medium–high	3,000	
Sedimentary rocks (e.g. cemented shale, siltstone, sandstone, limestone without cavities, thoroughly cemented conglomerates) in sound condition	Medium–high	1,000–4,000	
Compaction shale and other argillaceous rocks in sound condition	Low–medium	500–1,000	
Broken rocks of any kind with moderately close spacing of discontinuities (0.3 m or greater), except argillaceous rocks (e.g. shale)		1,000	
Limestone, sandstone, shale with closely spaced bedding		Assess in situ with load tests if necessary	
Heavily shattered or weathered rocks			

After Canadian Geotechnical Society (1992)

$$q_{ult}(\text{MPa}) = c(\sigma_c)^d \tag{9.4}$$

where $c = 4.8$ and $d = 0.5$. In most of the data used in suggesting this equation, the pile load has not really reached the ultimate value. Therefore, using $d = 0.5$ in Eq. (9.4) gives a conservative estimate of the ultimate end bearing pressure.

The *modulus ratio*, defined as the ratio of the Young's modulus (E) to the uniaxial compressive strength (σ_c), is a useful parameter for estimating E from σ_c. This ratio varies in the range of 100–1000, depending on the rock type. Typical values for the modulus ratios as suggested by Hoek and Diederichs (2006) are summarised in Table 9.8. The Young's modulus in the horizontal direction can be estimated as 75 % of the vertical modulus. This applies to soils and rocks.

Geophysical tests are quite common in rocks, where shear wave velocity (v_s) or compression wave (v_p) is measured. The wave velocities are related to the shear and bulk modulus as

Table 9.8 Typical values of modulus ratios

	Texture			
	Coarse	Medium	Fine	Very fine
Sedimentary	Conglomerates	Sandstones	Siltstones	Claystones
	300–400	200–350	350–400	200–300
	Breccias		Greywackes	Shales
	230–350		350	150–250^a
				Marls
				150–200
	Crystalline limestone	Sparitic limestone	Micritic limestone	Dolomite
	400–600	600–800	800–1000	350–500
		Gypsum	Anhydrite	Chalk
		(350)^c	(350)^c	1000+
Metamorphic	Marble	Hornfels	Quartzite	
	700–1000	400–700	300–450	
		Metasandstone		
		200–300		
	Migamatite	Amphibiolites	Gneiss	
	350–400	400–500	300–750^a	
		Schists	Phyllites/Mica Schist	Slates
		250–1100^a	300–800^a	400–600^a
Igneous	Granite^b	Diorite^b		
	300–550	300–350		
	Granodiorite			
	400–450			
	Gabro	Dolerite		
	400–500	300–400		
	Norite			
	350–400			
	Porphyries		Diabase	Peridotite
	(400)^c		300–350	250–300
		Rhyolite	Dacite	
		300–500	350–450	
		Andesite	Basalt	
		300–500	250–450	
	Agglomerate	Volcanic Breccia	Tuff	
	400–600	(500)^c	200–400	

After Hoek and Diederichs (2006)
^{a}Highly anisotropic rocks: the modulus ratio will be significantly different if normal strain and/or loading occurs parallel (high modulus ratio) or perpendicular (low modulus ratio) to a weakness plane. Uniaxial test loading direction should be equivalent to field application
^{b}Felsic Granitoids: Coarse grained or altered (High modulus ratio), fine grained (low modulus ratio)
^{c}No data available; Estimated on the basis of geological logic

$$v_s = \sqrt{\frac{G}{\rho}} \qquad (9.5)$$

$$v_p = \sqrt{\frac{K + \frac{4}{3}G}{\rho}} \qquad (9.6)$$

Noting that

$$G = \frac{E}{2(1+\nu)} \qquad (9.7)$$

and

$$K = \frac{E}{3(1-2\nu)} \qquad (9.8)$$

This can be used to estimate the small strain Young's modulus as

$$E = 2(1+\nu)\rho v_s^2 \qquad (9.9)$$

$$E = \frac{(1-2\nu)(1+\nu)}{(1-\nu)}\rho v_p^2 \qquad (9.10)$$

The intact rock specimens tested in the laboratory for UCS are free of joints and do not truly reflect the load-deformation behaviour of the larger rock mass. Rock mass rating (RMR) and tunnelling quality index (Q) and are two popular classification systems that were developed mainly for tunnelling in rocks. These ratings are assigned to the rock mass on the basis of the UCS of the intact rock, RQD, discontinuity spacing, orientation of the discontinuity, joint roughness, and ground water conditions. Geological strength index (GSI) was introduced more recently by Hoek (1994), again for classifying the rock mass on the basis of the discontinuities.

Tomlinson (2001) suggested that the rock mass, the Young's modulus can be determined as

$$E_M = j \times \text{modulus ratio} \times \sigma_c \qquad (9.11)$$

where, j is a mass factor that accounts for the discontinuity spacing and is given in Table 9.9. Some of the empirical expressions relating the rock mass modulus to the intact rock modulus and one of the three rock mass ratings are summarised below.

Coon and Merritt (1969) suggested that

$$E_M = E_R[0.0231(RQD) - 1.32] \qquad (9.12)$$

where E_R is the Young's modulus of the intact rock and E_M/E_R to be larger than 0.15. Bieniawski (1978) suggested that

Table 9.9 Mass factor j in Eq. (9.11)

Discontinuity spacing (mm)	<30	30–100	>100
Mass factor j	0.2	0.5	0.8

$$E_M = E_R(RQD/350) \quad \text{for RQD} < 70 \tag{9.13}$$

$$E_M = E_R\left[0.2 + \frac{RQD - 70}{37.5}\right] \quad \text{for RQD} > 70 \tag{9.14}$$

Kulhawy (1978) suggested that

$$E_M = E_R\left[0.1 + \frac{RMR}{1150 - 11.4RMR}\right] \quad \text{for RQD} > 70 \tag{9.15}$$

Serafim and Pereira (1983) suggested that, for $0 < RMR < 90$,

$$E_M(\text{GPa}) = 10^{\frac{RMR-10}{40}} \tag{9.16}$$

Equation (9.16) was later modified by Hoek et al. (2002) as

$$E_M(GPa) = \left(1 - \frac{D}{2}\right)\sqrt{\frac{\sigma_c(\text{MPa})}{100}} \times 10^{\frac{GSI-10}{40}} \quad \text{for } \sigma_c \leq 100 \text{ MPa} \tag{9.17}$$

$$E_M(GPa) = \left(1 - \frac{D}{2}\right) \times 10^{\frac{GSI-10}{40}} \quad \text{for } \sigma_c > 100 \text{ MPa} \tag{9.18}$$

Here, D is a factor to account for the disturbance in the rock mass due to blasting and stress relief, varying between 0 and 1; 0 for undisturbed and 1 for highly disturbed rock.

Bieniawski (1984) suggested that, for $45 < RMR < 90$

$$E_M(\text{GPa}) = 2RMR - 100 \tag{9.19}$$

Grimstad and Barton (1993) suggested that, for $1 < Q < 400$,

$$E_M(\text{GPa}) = 25\log Q \tag{9.20}$$

O'Neill et al. (1996) suggested the ratios of E_M/E_R given in Table 9.10.

9.4 Brazilian Indirect Tensile Strength

On rock samples, it is difficult to carry out a direct tensile strength test in the same way we test steel specimens. The main difficulties are in gripping the specimens without damaging them and applying stress concentrations at the loading grip, and

9.4 Brazilian Indirect Tensile Strength

Table 9.10 E_M/E_R values

RQD (%)	E_M/E_R	
	Closed joints	Open joints
100	1.00	0.60
70	0.70	0.10
50	0.15	0.10
20	0.05	0.05

After O'Neill et al. (1996)

in applying the load without eccentricity. *Indirect tensile strength test* (ASTM D3967; ISRM 1978b), also known as the *Brazilian test*, is an indirect way of measuring the tensile strength of a cylindrical rock specimen having the shape of a disc. The sample with thickness to diameter (t/d) ratio of 0.5 is subjected to a load that is spread over the entire thickness, applying a uniform vertical line load diametrically (Fig. 9.1). The load is increased to failure, where the sample generally splits along the vertical diametrical plane. From the theory of elasticity of an isotropic medium, the tensile strength of the rock σ_t is given by (Timoshenko 1934; Hondros 1959):

$$\sigma_t = \frac{2P}{dt} \tag{9.21}$$

where P = applied load, d = specimen diameter, and t = specimen thickness.

In the absence of any measurements, σ_t is sometimes assumed to be a small fraction of the uniaxial compressive strength σ_c. A wide range of values from 1/5 to 1/20 have been suggested in the literature, and 1/10 is a good first estimate. Some of the correlations between σ_c and σ_t are given in Table 9.11.

Sivakugan et al. (2014) showed that the cohesion and friction angle of intact rock can be estimated from Eqs. (9.22) and (9.23) given by

$$\phi = sin^{-1}\left(\frac{\sigma_c - 4\sigma_t}{\sigma_c - 2\sigma_t}\right) \tag{9.22}$$

$$c = \frac{0.5\sigma_c\sigma_t}{\sqrt{\sigma_t(\sigma_c - 3\sigma_t)}} \tag{9.23}$$

Here, it was assumed that the intact rock is isotropic and linearly elastic. It was also shown that

$$c = 1.82\sigma_t \tag{9.24}$$

Intact rock is often non-homogeneous and anisotropic. As a result, significant scatter is expected when using Eqs. (9.22), (9.23), and (9.24).

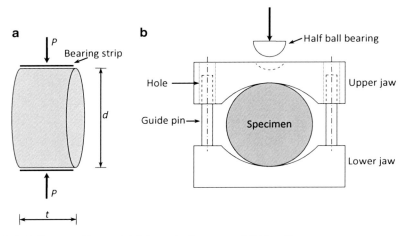

Fig. 9.1 Indirect tensile test: (**a**) Schematic diagram, and (**b**) Loading arrangement

Table 9.11 $\sigma_c - \sigma_t$ correlations

Correlation	Reference	Comments
$\sigma_c = 10.5\ \sigma_t + 1.2$	Hassani et al. (1979)	
$\sigma_c = 3.6\ \sigma_t + 15.2$	Szlavin (1974)	UK; 229 tests
$\sigma_c = 2.84\ \sigma_t - 3.34$	Hobbs (1964)	Mudstone, sandstone and limestone
$\sigma_c = 12.4\ \sigma_t - 9.0$	Gunsallus and Kulhawy (1984)	USA Dolostone, sandstone and limestone
$\sigma_c = 10\ \sigma_t$	Broch and Franklin (1972)	

9.5 Point Load Strength

The *point load test* (ASTM D5731; ISRM 1985; AS 4133.4.1) is an index test for strength classification of rocks, where a piece of rock is held between two conical platens of a portable light weight tester shown in Fig. 9.2. The load is increased to failure and the *point load index* $I_{s(50)}$ is calculated based on the failure load and the spacing between the cone tips. $I_{s(50)}$ is used to classify the rock and is roughly correlated to the strength parameters such as uniaxial compressive strength σ_c. The test is rather quick and can be conducted on regular rock cores or irregular rock fragments. A key advantage of point load test is that it can be carried out on an irregular rock fragment; this is not the case with most other tests where the specimens have to be machined and significant preparation is required. This makes it possible to do the tests at the site, on several samples in a relatively short time. Especially during the exploration stage, point load tests are very valuable in making informed decisions and can help in selecting the correct samples for the more sophisticated laboratory tests.

The test can also be used to quantify the strength anisotropy $I_{a(50)}$, the ratio of $I_{s\ (50)}$ in two perpendicular directions (e.g. horizontal and vertical). Historical

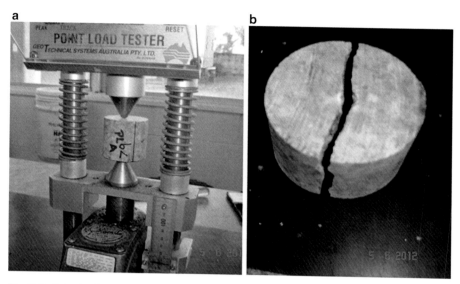

Fig. 9.2 Point load test: (**a**) Test equipment, and (**b**) Failed specimen

developments of the point load test and the theoretical background are discussed by Broch and Franklin (1972). $I_{s(50)}$ is related to σ_c by (Bieniawski 1975)

$$\sigma_c = 24\, I_{s(50)} \qquad (9.25)$$

The uniaxial compressive strength is the most used design parameter in rocks. The coefficient in Eq. (9.25) depends on the geology and the rock type. In the absence of UCS data, it may be conservatively be estimated as 20 $I_{s(50)}$. It is suggested that site specific correlations be developed or previously developed regional correlations be used.

9.6 Slake Durability

Rocks are generally weaker wet than dry, due to the presence of water in the cracks and its subsequent reaction to the applied loads during the tests. Repeated wetting and drying, which happens often in service, can weaken the rock significantly. *Slaking* is a process of disintegration of an aggregate when in contact with water. *Slake durability index* quantifies the resistance of a rock to wetting and drying cycles, and is seen as a measure of the *durability* of the rock. This is mainly used for weak rocks such as shales, mudstones, claystones and siltstones. The test procedures are described in ISRM (1979b), ASTM D4644 and AS 4133.3.4.

Figure 9.3 shows the slake durability apparatus which consists of two rotating sieve mesh drums immersed in a water bath. Ten rock lumps, each weighing

Fig. 9.3 Slake durability test apparatus

40–60 g, are placed in the drum and rotated for 10 min, allowing for disintegrated fragments to leave the drum through the 2 mm sieve mesh. The remaining fragments in the drum are dried and weighed. This is repeated over a second cycle of slaking, and the dry mass of the sample remaining in the drum, expressed as a percentage of the original mass in the drum at the beginning of the test, is known as the *second-cycle slake durability index* I_{d2} which varies in the range of 0-100 %. For samples that are highly susceptible to slaking I_{d2} is close to zero and for very durable rocks it is close to 100 %. A durability classification of rocks, based on slake durability index as proposed by Gamble (1971), is shown in Table 9.12. This table is slightly different to what is proposed by Franklin and Chandra (1972), who did not distinguish between the two different cycles and used a single durability index based on the first cycle. ASTM D4644 and ISRM (1979b) suggest reporting I_{d2} as the slake durability index.

9.7 Summary

The rock mass is made of intact rock and one or more sets of discontinuities. The intact rock specimens that are recovered from coring represent a relatively small volume of the rock mass and do not fully reflect the presence of discontinuities. The

Table 9.12 Durability classification based on slake durability index

Durability	I_{d1}	I_{d2}
Very high	>99	>98
High	98–99	95–98
Medium high	95–98	85–95
Medium	85–95	60–85
Low	60–85	30–60
Very low	<60	<30

Gamble (1971)

behaviour of the rock mass is governed mostly by the discontinuities than the properties of the intact rocks. Uniaxial compressive strength, Brazilian indirect tensile strength and the point load strength are some of the properties that reflect the strength of the intact rock. Typical values of this parameters and the interrelationships among them are discussed in this chapter.

References

AS 4133.3.4-2005 Method 3.4: Rock swelling and slake durability tests – determination of the slake durability index of rock samples, Australian Standard
AS 4133.4.1-2007 Method 4.1: Rock strength tests – determination of point load strength index, methods of testing rocks for engineering purposes, Australian Standard
AS 4133.4.2.1-2007 Method 4.2.1: Rock strength tests – determination of uniaxial compressive strength of 50 MPa and greater, methods of testing rocks for engineering purposes, Australian Standard
ASTM D3967-08 Standard test method for splitting tensile strength of intact rock core specimens
ASTM D4644-08 Standard test method for slake durability of shales and similar weak rocks
ASTM D5731-08 Standard test method for determination of the point load strength index and application to rock strength classifications
ASTM D7012-07e1 Standard test method for compressive strength and elastic moduli of intact rock core specimens under varying states of stress and temperatures
Bieniawski ZT (1975) The point load test in geotechnical practice. Eng Geol 9(1):1–11
Bieniawski ZT (1978) Determining rock mass deformability – experience from case histories. Int J Rock Mech Min Sci 15:237–247
Bieniawski ZT (1984) Rock mass design in mining and tunnelling. Balkema, Rotterdam, 272 p
Broch E, Franklin JA (1972) The point load strength test. Int J Rock Mech Min Sci 9(6):669–697
Canadian Geotechnical Society (1992) Canadian foundation engineering manual, 3rd edn. The Society, Vancouver, 511 pp
Coon RF, Merritt AT (1969) Predicting in situ modulus of deformation using rock quality, Special technical Publication No. 477. American Society for Testing Materials, Philadelphia
Deere DU (1964) Technical description of rock cores for engineering purposes. Rock Mech Eng Geol 1:17–22
Franklin JA, Chandra R (1972) The slake durability test. Int J Rock Mech Min Sci 9:325–341
Gamble JC (1971) Durability- plasticity classification of shales and other argillaceous rocks, PhD thesis, University of Illinois at Urbana-Champaign, IL, 159 pp
Gercek H (2007) Poisson's ratio values for rocks. Int J Rock Mech Min Sci 44(1):1–13

Goodman RE (1980) Introduction to rock mechanics. Wiley, New York

Gunsallus KL, Kulhawy FH (1984) A comparative evaluation of rock strength measures. Int J Rock Mech Min Sci Geomech Abstr 21(5):233–248

Grimstad E, Barton N (1993) Updating Q-system for NMT. In: Proceedings of the international symposium on sprayed concrete, Fagernes, Norway, Norwegian Concrete Association, Oslo, 20 pp

Handin J (1966) Strength and ductility. In: Clark SP (ed) Handbook of physical contacts. Geological Society of America, New York, pp 223–289

Hassani FP, Whittaker BN, Scoble MJ (1979) Strength characteristics of rocks associated with opencast coal mining in UK. In: Proceedings of 20th U.S. symposium on rock mechanics, Austin, Texas, pp 347–356

Hobbs DW (1964) Simple method for assessing uniaxial compressive strength of rock. Int J Rock Mech Min Sci Geomech Abstr 1(1):5–15

Hoek E (1994) Strength of rock and rock masses. ISRM News J 2(2):4–16

Hoek E, Brown ET (1997) Practical estimates of rock mass strength. Int J Rock Mech Min Sci Geomech Abstr 34(8):1165–1186

Hoek E, Diederichs MS (2006) Empirical estimation of rock mass modulus. Int J Rock Mech Min Sci 43:203–215

Hoek E, Carraza-Torres C, Corkum B (2002) Hoek-Brown failure criterion – 2002 edition. In: Proceedings of the 5th North American rock mechanics symposium, Toronto, Canada, 1, pp 263–273

Hondros G (1959) The evaluation of Poisson's ratio and the modulus of materials of a low tensile resistance by the Brazilian (indirect tensile) test with particular reference to concrete. Aust J Appl Sci 10(3):243–268

International Society of Rock Mechanics (ISRM), Commission on Standardisation of Laboratory and Field Tests (1978a) Suggested methods for the quantitative description of discontinuities in rock masses. Int J Rock Mech Min Sci Geomech Abstr 15(6):319–368

International Society of Rock Mechanics (ISRM), Commission on Standardisation of Laboratory and Field Tests (1978b) Suggested methods for determining tensile strength of rock materials. Int J Rock Mech Min Sci Geomech Abstr 15(3):99–103

International Society of Rock Mechanics (ISRM), Commission on Standardisation of Laboratory and Field Tests (1979a) Suggested methods for determination of the uniaxial compressive strength of rock materials. Int J Rock Mech Min Sci Geomech Abstr 16(2):135–140

International Society of Rock Mechanics (ISRM), Commission on Standardisation of Laboratory and Field Tests (1979b) Suggested methods for determining water content, porosity, density, absorption and related properties and swelling and slake durability index properties. Int J Rock Mech Min Sci Geomech Abstr 16(2):143–156

International Society of Rock Mechanics (ISRM), Commission on Standardisation of Laboratory and Field Tests (1985) Suggested method for determining point load strength. Int J Rock Mech Min Sci Geomech Abstr 22(2):51–60

Kulhawy FH (1978) Geomechanical model for rock foundation settlement. J Geotech Eng Div 104 (GT2):211–227

Lambe TW, Whitman RV (1979) Soil mechanics, SI version. Wiley, New York

O'Neill MW, Townsend FC, Hassan KH, Buller A, Chan PS (1996) Load transfer for drilled shafts in intermediate geomaterials, Report FHWA-RD-95-171. Federal Highway Administration, McLean, 184 p

Peck RB, Hanson WE, Thornburn TH (1974) Foundation Engineering, 2nd edn. Wiley, New York

Seidel JP, Haberfield CM (1994) Prediction of the variation of bored pile resistance with soil and rock strength, Aust Geomech J 26:57–64

Serafim JL (1968) Influence of interstitial water on the behavior of rock masses. In: Stagg KG, Zienkiewicz OC (eds) Rock mechanics in engineering practice. Wiley, London, pp 55–97

Serafim JL, Del Campo A (1965) Interstitial pressures on rock foundations of dams. J Soil Mech Found Div ASCE 91(SM5):66

References

Serafim JL, Pereira JP (1983) Considerations of the geomechanics classification of Bieniawski. In: Proceedings of the international symposium of engineering geology and underground construction, Lisbon, Portugal, pp 1133–1144

Sivakugan N, Das BM, Lovisa J, Patra CR (2014) Determination of c and ϕ of rocks from indirect tensile strength and uniaxial compression tests. Int J Geotech Eng 8(1), Maney Publishing, UK

Szlavin J (1974) Relationships between some physical properties of rocks determined by laboratory tests. Int J Rock Mech Min Sci 11:57–66

Timoshenko S (1934) Theory of elasticity. McGraw Hill, New York

Tomlinson MJ (2001) Foundation design and construction, 7th edn. Prentice Hall, Upper Saddle River

Zhang L, Einstein H (1998) End bearing capacity of drilled shafts in rock. J Geotech Eng ASCE 124(7):574–584

Index

A
Activity, 21, 22, 65
Air content, 14
Anisotropy, 29–30, 134, 203, 218
Area ratio, *See* Standard penetration test (SPT)
Atterberg limits
    liquid limit, 19–21
    plastic limit, 19–21
    shrinkage limit, 20

B
Bearing capacity, 78, 79, 81, 82, 107–108, 134, 139, 145–146, 165, 170, 174–176, 210, 213
Bjerrum correction, *See* Vane shear test
Blow count, *See* Penetration number/N-value
Bulk modulus, 40, 207, 213

C
Coefficient of
    consolidation, (*see* Consolidation)
    curvature, 18
    earth pressure at rest, 46, 79, 166, 186, 190
    gradation, 18
    secondary compression, (*see* Consolidation)
    uniformity, 17, 96
    volume compressibility, (*see* Consolidation)
Coefficient of earth pressure at rest, K_o, *See* Coefficient
Cohesion, 43, 44, 60–62, 103–104, 207, 217
Compaction
    maximum dry density, 22, 23
    modified proctor, 23, 24
    optimum water content, 22, 23
    standard proctor, 23
    zero air void curve, 23
Compression index, *See* Consolidation
Compression ratio/modified compression index, 57, 58
Cone penetration test (CPT), 4, 8, 77, 105, 107, 108, 115–117, 122, 128, 132, 134
    area ratio, net, 119
    corrected tip resistance, 119
    friction ratio, 122
    normalized friction ratio, 122
    pore pressure cone factor, 135
    pore pressure correction, 120
    pore water pressure parameter, 120
    sleeve friction, 118, 122, 141, 146
    tip resistance, 116, 118, 128, 134, 141, 145
Consistency index (CI), 20, 103
Consolidation
    coefficient of consolidation, 3, 35, 59, 117
    coefficient of secondary compression, 54
    coefficient of volume compressibility, 3, 34
    compression index, 34, 42, 54, 56, 57
    constant rate loading, 37, 39
    preconsolidation pressure, 34, 35, 47
    recompression index, 34, 35, 54, 57
    swelling index, 34, 57
Constant rate loading, *See* Consolidation
Constrained modulus, 35, 39, 58, 130–132, 137–138, 187, 189, 190
Core recovery ratio, 208
CPT, *See* Cone penetration test (CPT)
Critical hydraulic gradient, 32–33

D

Degree of consolidation, 36–38
Degree of saturation, 14, 23, 28, 72
Density
    bulk, 14, 16
    dry, 14, 16, 18, 22, 126
    saturated, 14
    submerged, 14
Density index, *See* Granular soil
Dilatancy angle, 46, 62, 63, 71
Dilatometer modulus, 186, 187, 189, 191
Dilatometer test, 5, 8, 183–191, 208
Discharge velocity, 28, 29
Drained loading, 44
Dynamic viscosity, *See* Permeability

E

Earthquake, 33, 106, 107, 148–150, 152
Effective grain size, 17, 53
Effective preconsolidation, 91
Effective stress, 32–34, 41, 44, 57, 60–62, 66, 81, 108, 130, 136, 152, 203
Empirical correlation, 5–8, 11, 28, 41, 45, 55, 58, 60, 117, 132, 203
Expansive clays, 21

F

Fines content, 149, 151–152
Friction angle
    critical state, 46, 65, 67, 71
    peak, 46, 63, 64, 66, 71, 129
    residual, 46, 64

G

Gap graded soils, *See* Granular soil
Grain size distribution, 16–18, 28, 67, 96
Granular soil
    dense, 17
    gap graded soil, 18
    loose, 17
    maximum void ratio, 18
    medium, 17
    minimum void ratio, 18, 96
    poorly graded soil, 18
    relative density, 16, 18, 46, 62, 66, 68, 70, 94, 96, 99, 125–129
    very dense, 127
    very loose, 127
    well graded soil, 18

H

Horizontal stress index, 186, 190, 191
Hydrometer, 17

I

Indirect tensile strength, 216–217
In situ lateral stress, 162, 165–167
In situ test, 115, 131, 160, 193
Intact rock, 8, 207–210, 215, 217

K

K_o, *See* Coefficient of earth pressure at rest

L

Laboratory test, 1, 8, 11, 12, 51–81, 198, 203
Laminar flow, *See* Permeability
LI, *See* Liquidity index (LI)
Liquefaction
    cyclic resistance ratio, 148
    cyclic shear strength, 148
    cyclic stress ratio, 106, 148
Liquidity index (LI), 20, 72, 73, 75, 76
Liquid limit, *See* Atterberg limits
Loading
    drained, 44
    undrained, 41, 44, 72

M

Material index, 186, 187, 191
Maximum dry density, *See* Compaction
Maximum void ratio, *See* Granular soil
Menard pressuremeter, *See* Pressuremeter
Minimum void ratios, *See* Granular soil
Modulus ratio, 77, 213, 214
Mohr-Coulomb, 43

N

N value, *See* Penetration number

O

OCR, *See* Over consolidation ratio (OCR)
Oedometer modulus, 35
Optimum water content, *See* Compaction
Over consolidation ratio (OCR), 74, 105, 127, 136, 186, 202

Index

P

Penetration number, 87, 91–99, 101, 106, 107
Permeability
  absolute permeability, 28
  discharge velocity, 28
  dynamic viscosity, 28, 29
  intrinsic permeability, 28
  laminar flow, 24–29
  seepage velocity, 27, 28
Permeability change index, 53
Phase diagram, 12, 13
Phase relations, 12–16, 32
PI, *See* Plasticity index (PI)
Piezocone, *See* Cone penetration test (CPT)
Pile designs, 81, 108
Plasticity, 18–21, 46, 63, 75, 108, 134, 135, 161, 198, 201
Plasticity index (PI), 20, 21, 54, 63–66, 74, 104, 135, 198, 201
Plastic limit, *See* Atterberg limits
Point load strength index, 207
Poisson's ratio, 40, 77–79, 207, 210, 211
Poorly graded soils, *See* Granular soil
Pore pressure parameters, 72
Pore water pressures, 33, 35–38, 41, 44, 72, 118, 120, 184, 186
Preconsolidation pressure, 34, 35, 58, 75, 104–105
Pressuremeter test
  limit pressure, 169, 175
  menard type, 160–161, 166, 167, 170, 173–177
  self-boring, 170

R

Recompression index, *See* Consolidation
Recompression ratio/modified recompression index, 57, 58
Relative density, *See* Granular soil
Reynold's number, 29
Rock mass, 8, 186, 207–209, 215, 216
Rock quality designation (RQD), 208

S

Secondary compression, 41–43, 60
Seismic cone, *See* Seismic piezocone
Seismic dilatometer, 191
Self-boring pressuremeter, *See* Pressuremeter

Sensitivity, 45, 72, 75, 76, 134, 135, 187, 193, 198, 199
Shear modulus, 41, 107, 117, 140, 167, 207
Shear strength
  critical state, 45–46, 71
  drained, 44
  peak, 45, 71
  residual, 46
  undrained, 44–45, 56, 73–76, 134, 165, 168–171, 186, 188, 194, 196–198, 201–203
Shear wave velocity, 107, 117, 132, 139, 141, 185, 191, 213
Shrinkage limit, *See* Atterberg limits
Sieve analysis, 52
Slake durability index, 219–221
Small strain shear modulus, 107, 132, 139, 185
Specific gravity (G_s), 14, 47
Split-spoon sampler, 87, 88, 183
Standard penetration test (SPT), 5, 8, 66–68, 73, 76, 87–108, 142, 159, 183
  area ratio, 88
  correction factor for hammer energy efficiency, 91–93
  correction factor for overburden, 151
Stiffness, 8, 22, 35, 59, 76–78, 106, 184, 207
Strengths, 169
Swelling index, *See* Consolidation

U

UCS, *See* Uniaxial compressive strength (UCS)
Undrained loading, 41, 44, 72
Uniaxial compressive strength (UCS), 210, 215, 217–219
Unified Soil Classification System (USCS), 20, 26
Unit weight
  bulk unit weight, 17
  dry unit weight, 14
  saturated unit weight, 17
  submerged unit weight, 14
USCS, *See* Unified Soil Classification System (USCS)

V

Vane shear test
  Bjerrum, 199
  correction factor, 198, 201, 203

V
Void ratio, 13, 14, 18, 22, 28, 33–35, 41, 42, 45, 46, 53, 54, 56, 57, 96, 126, 140

W
Water content, 13, 14, 20, 22, 29, 33, 44, 45, 54, 60, 61, 73, 207
Well graded soils, *See* Granular soil

Y
Young's modulus, *See* stiffness

Z
Zero air void curve, *See* Compaction